SOIL MECHANICS

SOIL MECHANICS

M. J. SMITH

M.Sc., C. Eng., M.I.C.E., M.I.Struct.E.
Senior Lecturer at the
South Bank University

FOURTH EDITION

 LONGMAN

Addison Wesley Longman Limited,
Edinburgh Gate, Harlow,
Essex CM20 2JE, England
and Associated Companies throughout the world.

First published 1967 by Macdonald & Evans Ltd
Reprinted 1968
Second edition 1970
Reprinted 1974, 1975
Third edition 1978
Reprinted 1979
Fourth edition published 1981 by George Godwin Ltd
Reprinted 1986
Reprinted by Longman Scientific & Technical 1988
Reprinted 1990, 1991, 1992, 1994
Reprinted by Addison Wesley Longman Limited 1998 (twice)

British Library Cataloguing in Publication Data

Smith, Michael John
  Soil Mechanics. – 4th ed. — (Godwin study guides).
  1. Soil mechanics
  2. Foundations
  I. Title
  624'.153   TA710

ISBN 0-582-03380-2

Transferred to digital print on demand 2002
Printed and bound by Antony Rowe Ltd, Eastbourne

GENERAL INTRODUCTION

This series was originally designed as an aid to students studying for technical examinations, the aim of each book being to provide a clear concise guide to the *basic principles* of the subject, reinforced by worked examples carefully selected to illustrate the text. The success of the series with students has justified the original aim, but it became apparent that qualified professional engineers in mid-career were finding the books useful.

In recognition of this need, the books in the series have been enlarged to cover a wider range of topics, whilst maintaining the concise form of presentation.

It is our belief that this increase in content should help students to see their study material in a more practical context without detracting from the value of the book as an aid to passing examinations. Equally, it is believed that the additional material will present a more complete picture to professional engineers of topics which they have not had occasion to use since completing their original studies.

A list of other books in the series is given at the front of this book. Further details may be obtained from the publishers.

M. J. Smith
General Editor

AUTHOR'S PREFACE

This book has been prepared with the object both of helping the student pass an examination, and for the practising engineer. The subject-matter has been restricted to basic principles, and worked examples are selected to give good coverage. The more descriptive aspects of soil mechanics, such as site investigation, piling practice and soil stabilisation have been omitted, and to obtain information on these subjects the reader is referred to the wealth of technical journals which deal with civil engineering contracts in which the latest practical techniques are described.

The value of soil mechanics to the civil engineer is often questioned, firstly, on the basis that empirical methods have been used for hundreds of years, and secondly, that soil is too complex a material to be subjected to scientific examination. Both of these reasons have some validity; indeed, it is dangerous to consider foundation problems on a scientific approach only without taking into account the knowledge of soil in practical conditions. However, the theory can be a great aid to understanding the practical case and to making more economic use of the material available. The variable nature of soil, even on a limited site, should also be considered when making an investigation, but if sufficient samples are taken, a good average value of soil properties may be obtained, or in certain cases the worst value may be the criterion; again experience is required in interpretation.

October 1980   M.J.S.

Tables from British Standard Code of Practice CP 2001: 1957 "Site Investigations", are reproduced by permission of the British Standards Institution, 2 Park Street, London, W.1, from whom copies of the complete code may be purchased.

CONTENTS

CHAPTER 1

SOIL DEFINITIONS

ENGINEERING DEFINITION OF SOIL

For engineering purposes, soil is considered to be any loose sedimentary deposit, such as gravel, sand, silt, clay or a mixture of these materials. It should not be confused with the *geological* definition of soil, which is the weathered organic material on the surface, or topsoil. Topsoil is generally removed before any engineering projects are carried out.

VOIDS

Soil is made up of various-sized particles packed together, with the spaces between particles known as *voids* (*see* Fig. 1(*a*)). The voids are generally a mixture of air and water, but in certain circumstances may be completely air or completely water.

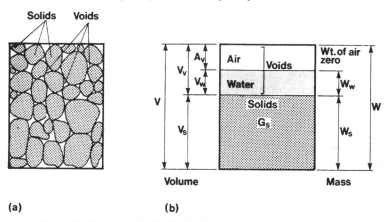

FIG. 1.(*a*) Soil sample, (*b*) Block diagram. Note symbols used.

Void ratio

The ratio of volume of voids to volume of solids is known as the *void ratio*.

$$\text{Void ratio } e = \frac{V_v}{V_s}$$

For convenience it may be assumed that all the solids in a sample can be compressed together and their volume considered equal to unit volume.

This may be shown in a block diagram (*see* Fig. 1(*b*)).

$$\text{If } V_s = 1$$

$$e = \frac{V_v}{1} = V_v$$

Total volume of soil sample $V = 1 + e$

Porosity

The porosity of a soil is defined as the ratio of volume of voids to total volume of sample.

$$\text{Porosity } n = \frac{V_v}{V}$$

which, referring to Fig. 1(*b*), taking $V_s = 1$ gives

$$\text{Porosity } n = \frac{e}{1 + e}$$

Degree of saturation

The ratio of volume of water to volume of voids is known as the *degree of saturation.*

$$\text{Degree of saturation } S_r = \frac{V_w}{V_v}$$

or　　　　　　　　Percentage saturation $= S_r \times 100$

The voids in soil below the water table should be considered as completely filled with water, in which case the degree of saturation is 1, or the percentage saturation 100 per cent. In fine-grained soils water will rise owing to capillary action between particles, and soil for a "fringe" height above the water table may become saturated. Even well above the water table there will always be a thin film of water surrounding individual grains of soil. This is known as *adsorbed* water. Adsorbed water can be removed only by oven-drying the soil, and is of importance in considering cohesion between grains.

Percentage air voids

The ratio of the volume of air to the total volume of soil is known as the percentage air voids.

$$\text{Percentage air voids } V_a = \frac{A_v}{V} \times 100$$

MOISTURE CONTENT, SPECIFIC GRAVITY AND DENSITY

The methods of determination of the specific gravity of soil particles, the moisture content and bulk density of a soil sample are given in detail in B.S. 1377: 1975 (*Methods of testing soils for civil engineering purposes*). Only a brief description therefore is given here, and for fuller details the British standard should be referred to.

These three properties of the soil should be determined in all site investigations and laboratory tests.

Moisture content of soil

Degree of saturation should not be confused with *moisture content*, which is the ratio of weight of water in the sample to weight of solids:

$$\text{Moisture content } m = \frac{W_w}{W_s}$$

or   Percentage moisture content $= m \times 100$

Determination of the moisture content of soil

A sample of soil is placed in a container of known weight, with a lid on to prevent evaporation. The container and soil are weighed and then placed in an oven at 105° C, with the lid removed, until the sample is dry. If the container, lid and dry soil are weighed again the loss in weight is the weight of water in the original sample, and the weight of solids is the final weight less the weight of the container. Hence the

$$\text{Moisture content } m = \frac{\text{Weight of water}}{\text{Weight of solids}}$$

may be determined.

Specific gravity of soil particles

The specific gravity of any material is defined as the ratio of the weight of a given volume of that material to the weight of an equal volume of water.

In a soil sample it is useful to know the specific gravity of the material of the soil particles. If this property G_s is known and the dry weight of the soil particles W_s known, then the volume of solids V_s may readily be determined, since:

$$\frac{W_s}{V_s} = G_s \gamma_w$$

where γ_w is the density of water (1000 kg/m^3).

For soil particles, which contain a high content of quartz, the specific gravity, G_s, is usually about 2·7.

Determination of the specific gravity of soil particles

To determine the specific gravity of soil particles a known weight of soil particles W_s (approximately 200 g of fine-grained or 400 g of coarse-grained soil) is thoroughly mixed with approximately 500 ml of water in a 1 litre jar. The jar is then filled to the top with water, the outside of the jar wiped dry and the jar plus soil plus water weighed, W_1. If the weight of the jar just filled with water is W_2, then:

Submerged weight of solids $= W_1 - W_2$

$$\text{Specific gravity of particles } G_s = \frac{\text{Weight of solid particles}}{\text{Weight of an equal volume of water}}$$

but

Weight of an equal volume of water
$=$ weight of water displaced by solids
$=$ weight of solids in air $-$ submerged weight of solids

$$\text{Hence specific gravity } G_s = \frac{W_s}{W_s - (W_1 - W_2)}$$

Bulk density of soil

The density of the complete soil sample (i.e. solids and voids) is usually expressed as *bulk* density.

$$\text{Bulk density } \gamma = \frac{W}{V}$$

Determination of bulk density

If a sample of soil can be taken in an undisturbed condition, then measurement of bulk density is simple. A cylindrical cutter about 100 mm diameter and 125 mm long is carefully driven into the soil, dug out, cleaned, trimmed and weighed. The weight of the cutter and its internal dimensions are readily determined, hence:

$$\text{Bulk density } \gamma = \frac{\text{Weight of cutter and soil} - \text{Weight of cutter}}{\text{Internal volume of cylinder}}$$

Obtaining an undisturbed sample is sometimes difficult, in which case a disturbed sample may be used. A hole is dug about 100 mm diameter and 150 mm deep and the excavated soil weighed. The volume of the hole may now be determined by filling it with a measured quantity of dry, uniformly graded sand of known density.

$$\text{Bulk density of soil } \gamma = \frac{\text{Weight of soil}}{\text{Volume of sand}}$$

In impervious soil, oil may be used instead of sand. The moisture content of the soil should be determined in each case.

Dry density
This is a special case of bulk density and is the density of the sample assuming the water is removed from the soil. The volume of the sample will not change, and therefore:

$$\text{Dry density } \gamma_d = \frac{W_s}{V}$$

The dry density is usually calculated from the measured values of bulk density and moisture content. The relationship between γ, γ_d and m is therefore of value, i.e.:

$$m = \frac{W_w}{W_s}$$

$$W = W_s + W_w$$
$$= W_s + mW_s$$
$$= W_s(1 + m)$$

also

$$\gamma = \frac{W}{V}$$
$$= \frac{W_s(1 + m)}{V}$$
$$= \gamma_d(1 + m)$$

or

$$\gamma_d = \frac{\gamma}{(1 + m)}$$

Saturated density
This is another special case of bulk density and is the density of the sample when the voids are completely filled with water. The volume of the sample will not change, and if the voids are filled with water the weight of this water $= V_v\gamma_w$

$$\text{Saturated density} = \frac{W_s + V_v\gamma_w}{V}$$

Submerged density
When the soil is below the water table it will be saturated, as previously noted, but it will also be submerged. Now:

Submerged density of soil

$$= \text{Bulk density of soil} - \text{Density of water}$$
$$(Archimedes'\ principle)$$

$$\gamma' = \gamma - \gamma_w$$

or, since the soil will be saturated,

$$\text{Submerged density } \gamma' = \gamma_{sat} - \gamma_w$$

The student, however, should avoid using submerged density as far as possible.

EXAMPLE 1

A sample of soil weighing 30·6 kg had a volume of 0·0183 m³. When dried out in an oven its weight was reduced to 27·2 kg. The specific gravity of the solids was found to be 2·65. Determine the following:

  (a) Bulk density.
  (b) Dry density.
  (c) Percentage moisture content.
  (d) Saturated density.
  (e) Percentage air voids.
  (f) Void ratio.
  (g) Porosity.
  (h) Degree of saturation.
  (i) Critical hydraulic gradient.

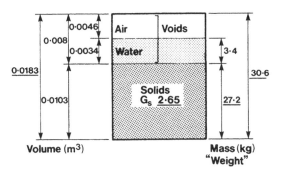

FIG. 2. Note that the figures underlined are given in the question

SOLUTION

This question may be shown in a block diagram (see Fig. 2).

  (a) Bulk density $\gamma = \dfrac{W}{V} = \dfrac{30\cdot6}{0\cdot0183} = 1672 \text{ kg/m}^3$

  (b) Dry density $\gamma_d = \dfrac{W_s}{V} = \dfrac{27\cdot2}{0\cdot0183} = 1486 \text{ kg/m}^3$

  (c) Weight of water in sample = 30·6 − 27·2 = 3·4 kg

$$\text{Moisture content } m = \frac{W_w}{W_s} = \frac{3\cdot4}{27\cdot2} = 0\cdot125$$

  or percentage moisture content = 12·5%

(d) Density of particles $\gamma_s = \dfrac{W_s}{V_s} = G_s\gamma_w$

$$\therefore \quad V_s = \frac{W_s}{G_s\gamma_w} = \frac{27\cdot2}{2\cdot65 \times 1000} = 0\cdot0103 \text{ m}^3$$

$$V_v = V - V_s = 0\cdot0183 - 0\cdot0103 = 0\cdot008 \text{ m}^3$$

If soil is saturated voids will be all water:

Saturated density $= \dfrac{W_s + V_v\gamma_w}{V} = \dfrac{27\cdot2 + 0\cdot008 \times 1000}{0\cdot0183} = \underline{\underline{1923 \text{ kg/m}^3}}$

(e) $V_w = \dfrac{3\cdot4}{1000} = 0\cdot0034 \text{ m}^3$

$A_v = 0\cdot008 - 0\cdot0034 = 0\cdot0046 \text{ m}^3$

Percentage air voids $V_a = \dfrac{A_v}{V} = \dfrac{0\cdot0046}{0\cdot0183} \times 100 = \underline{\underline{25\%}}$

(f) Void ratio $e = \dfrac{V_v}{V_s} = \dfrac{0\cdot008}{0\cdot0103} = \underline{\underline{0\cdot777}}$

(g) Porosity $n = \dfrac{V_v}{V} = \dfrac{0\cdot008}{0\cdot0183} = \underline{\underline{0\cdot437}}$

(*Note.* Porosity $= \dfrac{e}{1 + e} = \dfrac{0\cdot777}{1\cdot777} = 0\cdot437$)

(h) Degree of saturation $S_r = \dfrac{V_w}{V_v} = \dfrac{0\cdot0034}{0\cdot008} = 0\cdot425$

(j) Critical hydraulic gradient. This is discussed in Chapter 4 where an expression given is:

$$i_c = \frac{G_s - 1}{1 + e}$$

$\therefore$ Critical hydraulic gradient $= \dfrac{2\cdot65 - 1}{1 + 0\cdot777} = \underline{\underline{0\cdot93}}$

EXAMPLE 2

A laboratory test carried out on an undisturbed sample of soil weighing 1·74 kg and having a volume of $\frac{1}{1000}$ m^3 determined the specific gravity of the solids to be 2·6 and the dry density of the soil to be 1500 kg/m^3. Calculate:

(a) The moisture content.
(b) The void ratio and porosity.
(c) The critical hydraulic gradient.
(d) The saturated and submerged densities.
(e) The degree of saturation of the soil.

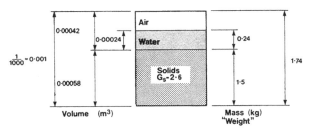

Fig. 3

SOLUTION
(see Fig. 3)

(a) $W_s = 1500 \times \frac{1}{1000} = 1.5$ kg

$W_w = 1.74 - 1.5 = 0.24$ kg

$$m = \frac{0.24}{1.5} = 0.16$$

or percentage moisture content = 16%

(b) $V_s = \dfrac{W_s}{G_s \gamma_w} = \dfrac{1.5}{2.6 \times 1000} = 0.00058$ m^3

$V_v = V - V_s = 0.001 - 0.00058 = 0.00042$ m^3

Void ratio $e = \dfrac{0.00042}{0.00058} = 0.72$

Porosity $n = \dfrac{V_v}{V} = \dfrac{0.00042}{0.001} = 0.42$

(c) $i_c = \dfrac{G_s - 1}{1 + e} = \dfrac{2.6 - 1}{1 + 0.72} = 0.93$

(d) $\gamma_{sat} = \dfrac{W_s + V_v \gamma_w}{V} = \dfrac{1.5 + 0.00042 \times 1000}{0.001} = 1920$ kg/m^3

$\gamma' = \gamma_{sat} - \gamma_w = 1920 - 1000 = 920$ kg/m^3

(e) $V_w = \dfrac{W_w}{\gamma_w} = \dfrac{0.24}{1000} = 0.00024$ m^3

$$S_r = \frac{V_w}{V_v} = \frac{0.00024}{0.00042} = 0.571$$

or percentage saturation = 57.1%

EXAMPLE 3

In order to measure the *in situ* density of a soil the following sand replacement test was carried out. 4·56 kg of soil were extracted from a hole at the surface of the soil. The hole was then just filled with 3·54 kg of loose dry sand.

(a) If it took 6·57 kg of the same sand to fill a container 0·0042 m³ in volume, determine the bulk density of the soil.

(b) In a water-content determination 24 g of the moist soil weighed 20 g after drying in an oven at 105° C. If the specific gravity of the particles was 2·68, determine the water content, the dry density and the degree of saturation of the soil.

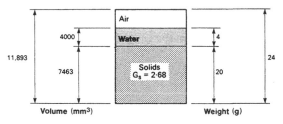

FIG. 4

SOLUTION

Referring to Fig. 4:

(a) Volume of hole $= \dfrac{0·0042}{6·57} \times 3·54 = 0·00226$ m³

Bulk density $\gamma = \dfrac{W}{V} = \dfrac{4·56}{0·00226} = \underline{\underline{2018 \text{ kg/m}^3}}$

(b) From moisture content determination:

$$m = \frac{W_w}{W_s} = \frac{4}{20} = 0·2$$

or percentage moisture content $= \underline{\underline{20\%}}$

Dry density $\gamma_d = \dfrac{\gamma}{1 + m} = \dfrac{2018}{1 + 0·2} = \underline{\underline{1681 \text{ kg/m}^3}}$

$$V = \frac{W}{\gamma} = \frac{24 \times 1000^3}{2018 \times 1000} = 11{,}893 \text{ m}^3$$

$$V_s = \frac{W_s}{G_s \gamma_w} = \frac{20 \times 1000^3}{2·68 \times 1000 \times 1000}$$

$$= 7463 \text{ mm}^3$$

$$V_v = V - V_s = 4430 \text{ mm}^3$$

$$V_w = 4000 \text{ mm}^3$$

$$S_r = \frac{4000}{4430} = 0·9$$

or percentage saturation $= \underline{\underline{90\%}}$

NEUTRAL AND EFFECTIVE STRESS

At any horizontal section, depth z in a soil profile, the total downward pressure is due to the weight of soil above the section.

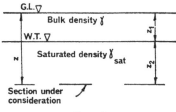

FIG. 5

Resistance to this pressure is provided, partly by the soil grains, and, if the section is below the water table, partly by the upward pressure of the water.

Total load at depth z per unit area $\sigma = z_1\gamma + z_2\gamma_{sat}$

This is resisted by the intergranular pressure σ', which is referred to as the *effective stress*, and by the upward water pressure u, which is referred to as the *neutral stress* and equals $z_2\gamma_w$, i.e.:

Total downward load per unit area = Intergranular pressure
+ Upward water pressure
= Effective stress
+ Neutral stress

$$\sigma = \sigma' + u$$

This relationship between load, effective and neutral stresses is of great importance in soil mechanics.

EXAMPLE 4
A bore hole on a building site has the soil profile shown in Fig. 6(a). Find the effective stress at the bottom of the clay:

(a) under normal conditions;
(b) if the ground water level is lowered 2·4 m by pumping (assume the sand remains saturated with capillary water up to the original level).

SOLUTION
Referring to Fig. 6(b):

$$\gamma_{sat} = \frac{W_s + V_v\gamma_w}{V} \qquad (\textit{see page 5})$$

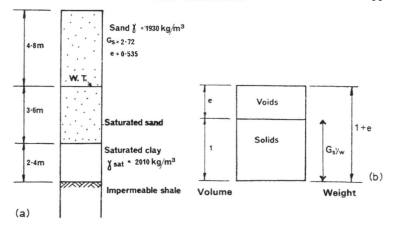

FIG. 6(a). Bore-hole log. (b) Block diagram.

Assuming $V_s = 1$ then $V_v = e$, $V = 1 + e$ and $W_s = G_s\gamma_w$

$$\therefore \quad \gamma_{sat} = \frac{G_s\gamma_w + e\gamma_w}{1 + e}$$

$$= \left(\frac{G_s + e}{1 + e}\right)\gamma_w$$

$\therefore$ for saturated sand

$$\gamma_{sat} = \frac{2\cdot72 + 0\cdot535}{1 + 0\cdot535} \times 1000 = \underline{\underline{2120 \text{ kg/m}^3}}$$

(a) Downward load/m² at base of clay

$$= (4\cdot8 \times 1930 + 3\cdot6 \times 2120 + 2\cdot4 \times 2010)9\cdot8/1000$$
$$= 213 \text{ kN/m}^2$$

Neutral stress $u = (3\cdot6 + 2\cdot4) \times 1000 \times 9\cdot8/1000 = 58\cdot8 \text{ kN/m}^2$

$$\sigma = \sigma' + u$$

$\therefore$ Effective stress $\sigma' = 213 - 58\cdot8 = \underline{\underline{154\cdot2 \text{ kN/m}^2}}$

Alternatively the submerged soil densities could be used, i.e.

$$\sigma' = [4\cdot8 \times 1930 + 3\cdot6(2120 - 1000)$$
$$+ 2\cdot4(2010 - 1000)]9\cdot8/1000$$

$$= \underline{\underline{154\cdot2 \text{ kN/m}^2}}$$

However the reader is recommended to use neutral stress for all calculations.

(b) When ground water level is lowered 2·4 m:

Downward load per m² at base of clay

$$\sigma = 213 \text{ kN/m}^2$$
$$\text{Neutral stress } u = (1·2 + 2·4)1000 \times 9·8/1000$$
$$= 35·3 \text{ kN/m}^2$$
$$\therefore \quad \text{Effective stress at base of clay} = 213 - 35·3 = \underline{\underline{177·7 \text{ kN/m}^2}}$$

QUESTIONS

1. A sample of saturated soil has a moisture content of 29 per cent and a bulk density of 1930 kg/m³. Determine the dry density and the void ratio of the soil and the specific gravity of the particles.

What would be the bulk density of a sample of this soil compacted to the same void ratio, but only 90 per cent saturated?

Note. Take the volume of the sample as 1 m³.

2. In a compaction test the weight of wet soil in the mould (volume $\frac{1}{1000}$ m³) was 1·88 kg. By drying out a small quantity of the soil its moisture content was found to be 20·7 per cent. The specific gravity of the particles was 2·72.

Find: (a) the dry density; (b) the void ratio; and (c) the percentage air voids.

If the sample was immersed in water and allowed to become completely saturated, without change in volume, calculate the saturated density and moisture content.

3. Derive an expression for the bulk density of partially saturated soil in terms of the specific gravity of the particles G_s, the void ratio e, the degree of saturation S_r and the density of water γ_w.

In a sample of clay the void ratio is 0·73 and the specific gravity of the particles is 2·71. If the voids are 92 per cent saturated, find the bulk density, the dry density and the percentage water content.

What would be the water content for complete saturation, the void ratio being the same.

Note. In this case take volume of solids as 1 m³.

4. A sample of soil, $\frac{1}{1000}$ m³ in volume, weighed in its natural state 1·73 kg, the degree of saturation being 61·6 per cent. After drying in the oven at 105° C the sample weighed 1·44 kg.

Find: (a) the specific gravity of the solids; (b) the natural water content; (c) the void ratio; (d) the bulk density, the dry density, the saturated density and the submerged density; (e) the critical hydraulic gradient.

5. A borehole log gave the following data:

0–2 m   Sand   Saturated density 1900 kg/m³
2–6 m   Silt   Saturated density 1800 kg/m³
6–9 m   Clay   Saturated density 2100 kg/m³

The water table is 4 m below the ground surface and the soil above the water table is saturated.

Calculate the effective pressure at the centre line of the clay.

(*i*) At the time of the investigation.

(*ii*) If the water table is lowered to the top of the clay.

(*iii*) If the water table is raised to the top of the silt.

(*iv*) If the water table is raised to ground level.

(*v*) If the water level rises 5 m above ground level.

Assume the soil remains saturated at all times.

CHAPTER 2

COMPACTION

In all civil engineering contracts soil is used as a construction material. Compaction of a soil is carried out to improve the soil properties and is a technique specified by the engineer. The most common examples are the sub-base of a road where the compaction may be *in situ*, or an embankment where the soil is brought in, frequently from a cutting being constructed elsewhere. Backfill after below ground construction is another example where soil may have to be compacted.

Compaction is carried out by rolling or tamping and causes compression of the soil by expelling air from the voids. It is not possible to remove water from the voids by compaction, but the addition of water to a slightly moist soil facilitates compaction by reducing surface tension. However, there is an optimum moisture content above which the addition of water causes an increase in voids.

The state of compaction is measured by the dry density (γ_d) where:

$$\gamma_d = \frac{W_s}{V}$$

but

$$\gamma = \frac{W}{V}$$

$$m = \frac{W_w}{W_s}$$

$$W = W_w + W_s$$
$$= mW_s + W_s$$
$$= W_s(1 + m)$$

$$\gamma = \frac{W_s(1 + m)}{V}$$

$$= \gamma_d(1 + m)$$

or

$$\gamma_d = \frac{\gamma}{1 + m}$$

Therefore to determine the compaction of a soil it is normal to find its bulk density and moisture content to be able to determine dry density.

PROCTOR TEST

The standard Proctor test is a method of finding the optimum moisture content for compaction of a soil. A cylindrical mould 0.001 m^3 in volume is filled with a soil sample in three layers, each layer being compacted by 27 blows of a standard hammer, weight 2.5 kg, length of drop 300 mm for each blow. (See B.S. 1377.)

The mould is then trimmed and weighed, hence giving the bulk density of the soil. The moisture content of the soil is then determined, and hence the dry density. The test is carried out with soil at different moisture contents and a graph of dry density against moisture content plotted. A heavy compaction test for soils subjected to greater compactive effort uses a 4.5 kg hammer dropping 450 mm on to five layers.

Examples of the curves obtained are shown in Fig. 7. From these curves the optimum moisture content can be read off at the point of maximum compaction (dry density).

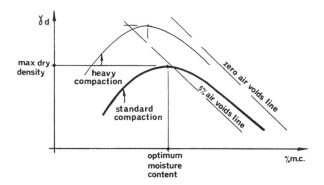

Fig. 7. Compaction curves: $\gamma_d \longrightarrow$ m.c. graph

Air void lines

For comparison a curve is plotted on the Proctor graph for the ideal case in which the soil is saturated and is known as the *zero air voids line*. Assuming the specific gravity of the soil particles G_s is known, a selection of values of moisture content around the optimum value are selected.

then $\gamma_d = \dfrac{W_s}{V}$ *and* $V_a = \dfrac{A_v}{V}$

Assuming unit volume of soil ($V = 1$)

$$V_a = A_v$$
$$= V - V_s - V_w$$
$$= 1 - \frac{W_s}{G_s \gamma_w} - \frac{W_w}{\gamma_w}$$

or

$$(1 - V_a)G_s \gamma_w = W_s + W_w G_s$$
$$= W_s(1 + mG_s)$$
$$W_s = \frac{G_s \gamma_w (1 - V_a)}{1 + mG_s}$$

but

$$\gamma_d = \frac{W_s}{V} = W_s$$

$$\therefore \quad \gamma_d = \frac{G_s \gamma_w (1 - V_a)}{1 + mG_s} \qquad \text{Eqn} \quad 1$$

for a saturated soil $V_a = 0$ and $\gamma_d = \dfrac{G_s \gamma_w}{1 + mG_s}$

However, complete saturation is impossible to obtain by compaction, and there will be some air voids in the compacted sample. A measure of the air voids can be seen at a glance if a 5 per cent and 10 per cent air voids line is also plotted on the Proctor graph. Equation 1 is therefore used to plot the 5 per cent and 10 per cent air voids line as well as the zero air voids line.

EXAMPLE 5
Standard Proctor compaction tests carried out on a sample of sandy clay yielded the following results:

Bulk density (kg/m³):	2058	2125	2152	2159	2140
Moisture content (%)	12·9	14·3	15·7	16·9	17·9

(a) Plot the curve of dry density against moisture content and hence find the maximum dry density and the optimum moisture content.

(b) Calculate the moisture content necessary for complete saturation at this maximum dry density if the specific gravity of the solid constituents is 2·73.

(c) Plot the "zero air voids" line and 5 per cent air voids line.

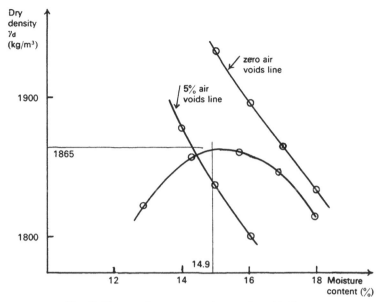

FIG. 8. Compaction curve and zero air voids line.

SOLUTION

(a) Referring to Fig. 8:

$$\gamma_d = \frac{\gamma}{1 + m}$$

$m =$	0·129	0·143	0·157	0·169	0·179
$\gamma =$	2058	2125	2152	2159	2140
$\gamma_d =$	1823	1859	1860	1846	1815

From graph   maximum dry density = **1865 kg/m³**

and   optimum moisture content = **14·9%**

(b) When dry density = 1865 kg/m³

For 1 m³ sample:

$$\text{Volume of solids} = \frac{1865}{2\cdot73 \times 1000} = 0\cdot683 \text{ m}^3$$

$$\text{Volume of water for saturation} = 0\cdot317 \text{ m}^3$$

$$\text{Weight of water} = 0\cdot317 \times 1000 = 317 \text{ kg}$$

$$\text{Moisture content} = \frac{317}{1865} \times 100 = \underline{\underline{17\%}}$$

(c) Select values of moisture content

$m(\%)$	14	15	16	17	18
$1 + mG_s$	1·38	1·41	1·44	1·46	1·49

for zero air voids $1 - V_a = 1$ $G_s\gamma_w = 2730$

γ_d	1978	1936	1896	1870	1832

for 5% air voids $1 - V_a = 0.95$ $G_s\gamma_w(1 - V_a) = 2594$

γ_d	1880	1840	1801	1777	1741

These lines are plotted on the graph Fig. 8.

SITE PROCEDURE

The Proctor test should be considered merely as a guide to compaction on site. The optimum moisture content should not be specified, since climatic conditions must be taken into account as well as the difficulties of accurately controlling moisture content on site. It is more common to specify a relative compaction for the site where:

$$\text{Relative compaction} = \frac{\text{Site value of dry density}}{\text{Proctor maximum dry density}} \times 100$$

A relative compaction of 90–95 per cent is commonly required, but this will depend on the contract and plant used.

There are many types of rollers used, i.e. rubber wheeled, smooth rollers, vibrating rollers all of varying weights. Selection of the type of roller will depend largely on the type of soil being compacted. Although a heavier roller will give better compaction, it is usually preferable to select the weight of roller and improve compaction by increasing the number of passes it makes over the soil. The contractor will normally wish to keep the number of passes to a minimum.

For an embankment the best procedure is to compact a trial area and measure the dry density of this area. When the relative compaction is satisfactory, then the number of passes of the roller used to construct the trial area is specified for the actual embankment.

The soil is normally compacted in layers of 200–300 mm, and a constant check made to ensure compaction is satisfactory.

In an embankment the rate of construction is also controlled due to the build up of pore-water pressure, but this is outside the scope of this chapter.

QUESTIONS

1. The results of a standard Proctor compaction test are as follows:

Moisture content m%	5	8	9	11	12	15	20
Bulk density γ (kg/m^3)	1890	2139	2170	2210	2219	2161	2069

Plot the curve of moisture content against dry density and determine the optimum moisture content and maximum compaction.

If the grain specific gravity is 2·70 plot the zero air voids and 10 per cent air voids line.

What are the values of void ratio, porosity and degree of saturation for the soil at its condition of optimum moisture content?

2. (a) Describe the standard Proctor compaction test.

(b) How can the effectiveness of compaction in the field be assessed?

(c) In a standard compaction test on a soil the following results were obtained:

Bulk density kg/m^3	2070	2139	2187	2212	2228	2211	2193
Moisture content %	6·8	8·5	9·4	10·2	11·3	12·5	13·6

What is the optimum moisture content and maximum compaction?

(d) If the specific gravity of the solids is 2·65 plot the zero air voids line.

(e) Find the percentage air voids at the maximum dry density and optimum moisture content.

(f) Plot the percentage air voids line for the air voids value found in part (e).

(g) Determine the percentage moisture content required to saturate the soil at its maximum dry density.

3. The following results were obtained from a standard Proctor compaction test on a cylindrical mould of volume 0·001 m^3.

Weight of sample of wet soil (g)	6·65	6·12	5·02	5·18	5·20	4·77	4·74
Weight of sample of dry soil (g)	6·03	5·51	4·49	4·60	4·59	4·18	4·12
Weight of soil in mould after compaction (g)	2821	2864	2904	2906	2895	2874	2834

Weight of mould = 1034 g

Assuming a relative compaction on site of 96 per cent at optimum moisture content, estimate the total pressure at the base of a 20 m high embankment built of this soil.

What will be the void ratio and degree of saturation of the soil in this embankment if the grain specific gravity is 2·67?

CHAPTER 3

SOIL CLASSIFICATION

CLASSIFICATIONS USED IN THIS CHAPTER

As stated at the beginning of Chapter 1, in civil engineering a soil may be taken to include any loose sedimentary deposit. This could include sand, clay, gravel, marl, etc. In order to be able to discuss the properties of different types of soil it is first necessary to have some way of classifying them. There are several ways in which a soil may be classified: by geological origin, by mineral content, by grain size or by plasticity. The last two are most widely used by engineers, and will be dealt with in this chapter.

FIELD IDENTIFICATION

For field identification of soils the Building Research Station has adopted a simple series of tests and this table is reproduced here (*see* Table 2).

If you were responsible for carrying out preliminary site investigations you could prepare a scheme of simple tests for examining and identifying any soil you might encounter. The tests should require no special apparatus, and their purpose should be:

(*a*) To distinguish between the main soil types.
(*b*) To assess the strength and structure of the soil.

Grain size. In this system, soils are split into coarse-grained non-cohesive, fine-grained cohesive and organic soils. They are then further subdivided into gravels, sands, silts, etc. The division of the coarse-grained non-cohesive soils into gravels and sands is according to grain size, which is readily determined by sieving.

Plasticity. The fine-grained cohesive soils are divided into silts and clays according to their plasticity. In the field, plasticity is determined by touch. Clays cannot be powdered when dry and are sticky when wet. Silts, on the other hand, are readily powdered when dry, and exhibit marked dilatancy when wet, that is, the moisture at the surface will recede if pressure is applied. Most students will have observed this phenomenon if they have stepped on damp fine sand on the beach and noticed the "dry" patch which forms around the foot. Most soils are a mixture of the various types, and a few examples of composite types are given in column 3.

The tests to assess the strength of the soil give in column 4 are very easy to apply, and the structure in column 5 is determined by simple visual examination.

It should be remembered that this classification is merely for use in the field. More elaborate tests are required for complete classification and for determining the strength of a soil.

PARTICLE SIZE DISTRIBUTION

Most systems of soil classification depend to some extent upon the distribution of various-sized particles in the soil. For coarse-grained material this distribution may be determined by sieving, and for finer particles a method of measuring the rate of settlement in water is used. The determination of particle-sized distribution by these methods is known as *mechanical* analysis.

Several systems of particle-size classification are in use, but the British Standards Institution had adopted that evolved by the Massachusetts Institute of Technology, since the boundaries of the main divisions correspond, approximately, to important changes in the engineering properties of the soil. These boundaries, together with a detailed description of the tests, are given in B.S. 1377: 1975, and therefore only a brief description is given here (see Table 1).

TABLE 1. PARTICLE SIZE LIMITS

Type	Range of particle size, mm
Cobbles	200–60
Coarse gravel	60–20
Medium gravel	20–6
Fine gravel	6–2
Coarse sand	2–0·6
Medium sand	0·6–0·2
Fine sand	0·2–0·06
Coarse silt	0·06–0·02
Medium silt	0·02–0·006
Fine silt	0·006–0·002
Clay	Less than 0·002

Coarse analysis (sieve test)

For coarse analysis either wet or dry sieving may be used. In either case an oven-dried sample of soil is weighed and passed through a batch of sieves.

TABLE 2. GENERAL BASIS FOR FIELD

			Size and nature of particles	Composite types 3
			Principal soil types	
			1 2	
		Types	Field identification	
Coarse grained, non-cohesive		Boulders Cobbles	Larger than 200 mm in diameter Mostly between 200 mm and 80 mm	Boulder gravels Hoggin
		Gravels	Mostly between 80 mm and 2 mm sieve	Sandy gravels
	Sands	Uniform	Composed of particles mostly between 2 mm and 63 μm sieves, and visible to the naked eye. Very little or no cohesion when dry Sands may be classified as uniform or well graded according to the distribution of particle size Uniform sands may be divided into coarse sands between 2 mm and 0·5 mm sieves, medium sands between 0·5 mm and 0·25 mm sieves and fine sands between 0·25 and 63 μm sieves	Silty sands Micaceous sands Lateritic sands
		Graded		Clayey sands
Fine grained, cohesive	Silts	Low Plasticity	Particles mostly passing 63 μm sieve Particles mostly invisible or barely visible to the naked eye. Some plasticity and exhibits marked dilatancy. Dries moderately quickly and can be dusted off the fingers. Dry lumps possess cohesion, but can be powdered easily in the fingers	Loams Clayey silts Organic silts Micaceous silts
	Clays	Medium Plasticity	Dry lumps can be broken but not powdered. They also disintegrate under water Smooth touch and plastic, no dilatancy. Sticks to the fingers and dries slowly Shrinks appreciably on drying, usually showing cracks Lean and fat clays show those properties to a moderate and high degree respectively	Boulder clays Sandy clays Silty clays
		High Plasticity		Marls Organic clays Lateritic clays
Organic		Peats	Fibrous organic material, usually brown or black in colour	Sandy, silty or clayey peats

IDENTIFICATION AND CLASSIFICATION OF SOILS

	Strength and structural characteristics			
	Strength 4		Strength 5	
Term	Field test	Term	Field identification	
Loose	Can be excavated with spade. 50 mm wooden peg can be easily driven	Homo-geneous	Deposit consisting essentially of one type	
Compact	Requires pick for excavation. 50 mm wooden peg hard to drive more than a few inches			
Slightly cemented	Visual examination Pick removes soil in lumps which can be abraded with thumb	Stratified	Alternating layers of varying types	
Soft	Easily moulded in the fingers	Homo-geneous	Deposit consisting essentially of one type	
Firm	Can be moulded by strong pressure in the fingers	Stratified	Alternating layers of varying types	
Very soft	Exudes between fingers when squeezed in fist	Fissured	Breaks into polyhedral fragments along fissure planes	
Soft	Easily moulded in fingers	Intact	No fissures	
Firm	Can be moulded by strong pressure in the fingers	Homo-geneous	Deposits consisting essentially of one type	
		Stratified	Alternating layers of varying types. If layers are thin the soil may be described as laminated	
Stiff	Cannot be moulded in fingers			
Hard	Brittle or very tough	Weathered	Usually exhibits crumb or columnar structure	
Firm	Fibres compressed together			
Spongy	Very compressible and open structure			

The weight of dry soil retained on each sieve is recorded, and the percentage of the total sample passing each of the sieves is calculated. This percentage passing is plotted on the sand and gravel fractions of a semi-logarithmic chart as shown in Fig. 9. The silt and clay fractions of the chart are completed after fine analysis of the soil.

Fine analysis

The theory of fine analysis is based on Stokes's law of settlement, i.e. small spheres in a liquid settle at different rates according to the size of the sphere.

The terminal velocity of a spherical soil particle settling in water is expressed by Stokes's law as:

$$v = \frac{\gamma_s - \gamma_w}{18\mu_w}D^2$$

where: γ_s = Density of soil particle;
γ_w = Density of water;
μ_w = Viscosity of water;
D = Diameter of the spherical particle.

For soil an average figure for γ_s is 2670 kg/m³. The density of water γ_w is 1000 kg/m³ and the viscosity of water at varying temperatures μ_w can be found from tables. At 20° C the viscosity is 0·001009 Ns/m², which gives:

$$V = \frac{(2670 - 1000) \times 9·81}{18 \times 0·001}D^2 \text{ m/s}$$

with D in metre units.

or $V \simeq \underline{\underline{900 \ D^2 \text{ mm/s}}}$

with D in mm units.

In practice, soil particles are never truly spherical. To overcome this, particle size is defined in terms of *equivalent diameter*, where the equivalent diameter of a particle is the diameter of an imaginary sphere of the same material which would sink in water with the same velocity as the irregular particle in question. It is this equivalent diameter, therefore, which is finally determined.

Stokes's law should only be applied to spheres between about 0·2 and 0·0002 mm in diameter. This, however, is within the limits required for classification of silt particles.

Experimental procedure for fine analysis
The soil must first be pre-treated to remove organic matter, the weight of which should be recorded. This is a lengthy and pains-

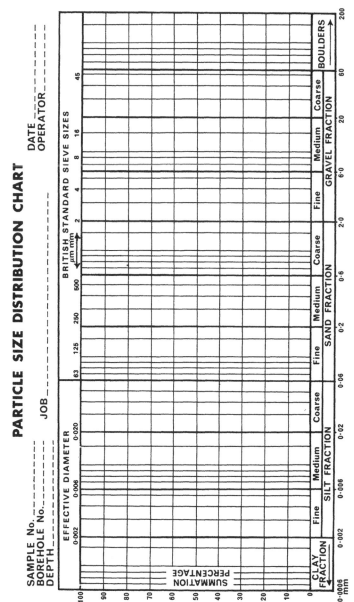

Fig. 9. Particle-size distribution chart

taking process, described in detail in B.S. 1377. A dispersing agent is now added to prevent flocculation, and coarse particles are removed by washing through a 63 μm sieve. The material retained on the sieve is dried and treated as for sands and gravels. The washing water is then subjected to sedimentation.

There are two methods of sedimentation analysis: (a) the pipette method, and (b) the hydrometer method.

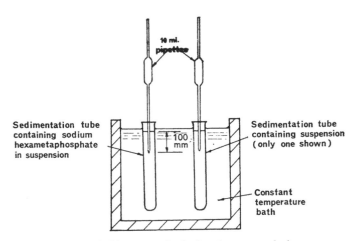

FIG. 10. Pipette method of settlement analysis

(a) *Pipette method.* The washing water containing the fine particles (weight W_b) is made up to 500 ml with distilled water and placed in a constant-temperature bath (*see* Fig. 10). When the suspension has reached the temperature of the bath it is taken out, shaken to disperse the particles and replaced in the bath. A stop watch is started immediately the suspension is replaced.

After a given time t (3–4 minutes depending on the specific gravity of the sample) a 10-ml sample of suspension is taken by pipette from a depth of 100 mm and the weight of solids in this sample found (W_D).

A correction must be made for the weight of dispersing agent (sodium hexametaphosphate) in the suspension. To do this a separate solution of the dispersing agent is tested at the same time in the same manner.

The whole procedure is repeated after 40–50 minutes and again after a further 6–7 hours.

From Stokes's law the velocity of the particles $v = KD^2$, where K is a constant equal to $\dfrac{\gamma_s - \gamma_w}{18\mu_w}$.

After time t_1 all the particles of a certain size D_1 will have settled from the surface to a depth of 100 mm. Any particles larger than size D_1 will have sunk below the 100 mm mark in the suspension. The velocity of particles size D_1 can therefore be calculated since they have moved a distance of 100 mm in known time t_1, i.e.:

$$v = \frac{h}{t_1}$$

$$KD_1{}^2 = \frac{h}{t_1}$$

$$D_1{}^2 = \frac{h}{Kt_1}$$

As h, t_1 and K are known, the maximum grain size D_1, at depth 100 mm after time t_1, can be calculated.

Since all sizes smaller than D_1 at this depth will be present in the same concentration as they were in the original suspension (see Fig. 11), percentage of particles less than size D_1 in original solution

$$N_1 = \frac{\text{Weight of solids per ml at depth 100 mm after time } t}{\text{Weight of solids per ml in original suspension}} \times 100$$

$$= \frac{W_{D1}/10}{W_b/500} \times 100$$

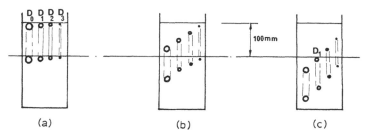

(a)   (b)   (c)

FIG. 11. Particle concentration on settlement. (a) Original suspension. Particles are completely dispersed and hence all particles sizes will be present at any horizontal section. (b) Intermediate suspension. Larger particles settle faster. (c) Suspension after time t_1. All particles larger than size D_1 have sunk below the 100 mm mark. Smaller particles are in the same concentration. Note that for convenience it is assumed that particles of the same size lie above each other.

Values for D_2 and D_3 can be found in time t_2 and t_3. These values can be plotted in the silt and clay fraction of the particle-size distribution chart.

(b) *Hydrometer method.* A 1000 ml suspension of fine particles is prepared in a similar manner to that in the pipette method, and the specific gravity of the suspension at depth h is measured at given intervals of time, using a hydrometer (*see* Fig. 12).

The hydrometer gives a direct reading for the specific gravity of the suspension. For convenience, the 1 is often omitted from the specific gravity reading on hydrometers and the decimal point moved three places to the right, i.e., a reading of 12 on the hydrometer means the specific gravity of the suspension is 1·012.

Assuming units are grammes and millilitres, then $\gamma_w = 1$.

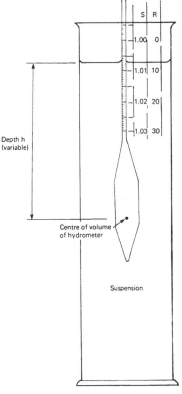

FIG. 12. Hydrometer method of settlement analysis. S = specific gravity of suspension. R = reading on hydrometer. $R = (S - 1)1000$.

In original suspension of 1000 ml:

$$\text{Weight of solids} = W_b$$

$$\therefore \quad \text{Volume of solids } V_b = \frac{W_b}{G_s \gamma_w}$$

$$\therefore \quad \text{Volume of water} = 1000 - \frac{W_b}{G_s \gamma_w}$$

$$\therefore \quad \text{Weight of water} = 1000\gamma_w - \frac{W_b}{G_s}$$

$$\text{Initial density of suspension } \gamma_i = \frac{W_b + 1000\gamma_w - W_b/G_s}{1000}$$

$$\gamma_i = \gamma_w + \frac{W_b}{1000}\left(\frac{G_s - 1}{G_s}\right)$$

and density of suspension at depth h after time t_1

$$= \gamma_t = \gamma_w + \frac{W_D}{10000}\left(\frac{G_s - 1}{G_s}\right).$$

Percentage of particles less than size D_1 in original suspension $= N_1$

$$N_1 = \frac{W_D}{W_b} \times 100$$

$$\therefore \quad \gamma_t = \gamma_w + \frac{N_1 W_b}{100\,000}\left(\frac{G_s - 1}{G_s}\right)$$

$$N_1 = (\gamma_t - \gamma_w)\frac{1000\,000\,G_s}{W_b(G_s - 1)}$$

$$N_1 = \frac{100\,000\,G_s\gamma_w}{W_b(G_s - 1)}(S - 1)$$

$$= \frac{100\,000\,G_s\gamma_w}{W_b(G_s - 1)}\frac{R}{1000}$$

but $\gamma_w = 1$

$$\therefore \quad N_1 = \frac{100\,G_s R}{W_b(G_s - 1)}$$

Readings of the hydrometer should be taken after $\frac{1}{2}$, 1, 2, 4, 8, 15 and 30 minutes, 1, 2 and 4 hours, and then once or twice daily.

Corrections must be made for temperature variations and the addition of dispersing agent. It should also be noticed that as the hydrometer sinks, h will increase slightly. It is therefore normal to calibrate the hydrometer before use.

EXAMPLE 6

In a sedimentation test 20 g of soil of specific gravity 2·69 and passing a 63 μm sieve were dispersed in 1000 ml of water having a viscosity of 0·001 SI units. One hour after the commencement of sedimentation, 20 ml of the suspension were taken by means of a pipette from a depth of 100 mm. The amount of solid particles (in the sample of 20 ml taken by pipette) obtained on drying was 0·07 g. Compute the following:

(a) The largest size of particle remaining in suspension at a depth of 100 mm, 1 hour after the commencement of sedimentation.

(b) The percentage of particles finer than this size in the original sample.

(c) The time interval from the commencement, after which the largest particle remaining in suspension at 100 mm depth is one-quarter of this size.

SOLUTION

(a) $v = \dfrac{\gamma_s - \gamma_w}{18\mu_w}D^2 = \dfrac{(2690 - 1000) \times 9\cdot81}{18 \times 0\cdot001}D^2$ m/s

also $v = \dfrac{h}{t} = \dfrac{100}{60 \times 60}$ mm/s

$$D = \sqrt{\dfrac{100 \times 18 \times 0\cdot001}{60 \times 60 \times 1000 \times 1690 \times 9\cdot81}}\ \text{m}$$

$$= \underline{\underline{0\cdot0055\ \text{mm}}}$$

(b) $W_b = 20$ g

at commencement volume of solids $= \dfrac{20}{2\cdot69} = 7\cdot4$ ml

∴ volume of solution $V_{sol} = 1007\cdot5$ ml

after 1 hour $W_D = 0\cdot07$ g

percentage particles less than 0.0055 mm $= \dfrac{W_D/20}{W_b/V_{sol}} \times 100$

$$= \dfrac{0\cdot07/20}{20/1007\cdot5} \times 100 = \underline{\underline{17\cdot6\%}}$$

(c) $v = KD^2 = \dfrac{h}{t}$

∴ if D is multiplied by $\frac{1}{4}$, then D^2 must be multiplied by $\frac{1}{16}$ and t must be multiplied by 16, i.e. $\underline{\underline{16\ \text{hours}}}$

EXAMPLE 7

(a) A sample of soil weighing 50 g is dispersed in 1000 ml of water. How long after the commencement of sedimentation should the hydrometer reading be taken in order to estimate the percentage of particles less than 0·002 mm effective diameter, if the centre of the hydrometer is 150 mm below the surface of the water?

$G_s = 2\cdot7$ Viscosity of water $\mu = 0\cdot001$ SI units

SOLUTION
From Stokes's law

$$v = \dfrac{(2700 - 1000) \times 9\cdot81}{18 \times 0\cdot001}\left(\dfrac{0\cdot002}{1000}\right)^2 \times 1000$$

$$= 0\cdot0037\ \text{mm/s}$$

Time of reading $= \dfrac{150}{0\cdot0037 \times 60 \times 60} = \underline{\underline{11\cdot26\ \text{hr}}}$

Particle-size distribution curve
Coarse and fine analysis having been carried out, the particle-size distribution curve may be plotted. The soil can be described according to the shape of the curve and where it fits on to the chart.

A uniform soil, where all the particles are approximately the same size, will have an almost vertical curve. A well-graded soil, containing a wide range of particle size, will have a curve spread evenly across the chart. A poorly graded soil will stretch across the chart but be deficient in intermediate sizes. Some examples are shown in Fig. 13.

Effective size
This is defined as the maximum particle size of the smallest 10 per cent and is denoted as D_{10}, i.e. for curves shown in Fig. 13:

Curve 1, $D_{10} = 0.006$ mm
Curve 2, $D_{10} = 0.052$ mm
Curve 3, $D_{10} = 0.07$ mm

Allen Hazen's uniformity coefficient
This is the ratio of the maximum particle size of the smallest 60 per cent to the effective size, and is denoted as U.

$$U = \frac{D_{60}}{D_{10}}$$

A uniform soil will have a coefficient approaching 1, whereas a well graded soil will have a high uniformity coefficient, i.e. for curves in Fig. 13:

$$\text{Curve 1, } U = \frac{0.044}{0.006} = 7.33$$

$$\text{Curve 2, } U = \frac{0.11}{0.052} = 2.12$$

$$\text{Curve 3, } U = \frac{0.4}{0.07} = 5.71$$

Thus uniformity coefficient and effective size give two points on the curve, which is often sufficient to define the curve.

EXAMPLE 8
The results of a sieving analysis of a soil were as follows:

Retained on sieve size (mm)	Weight retained (g)	Retained on sieve size (mm)	Weight retained (g)
20	0	2	3·5
12·5	1·7	1·4	1·1
10	2·3	0·5	30·5
6·3	8·4	0·355	45·3
5·6	5·7	0·180	25·4
2·8	12·9	0·063	7·4

The total weight of the sample was 147·2 g.

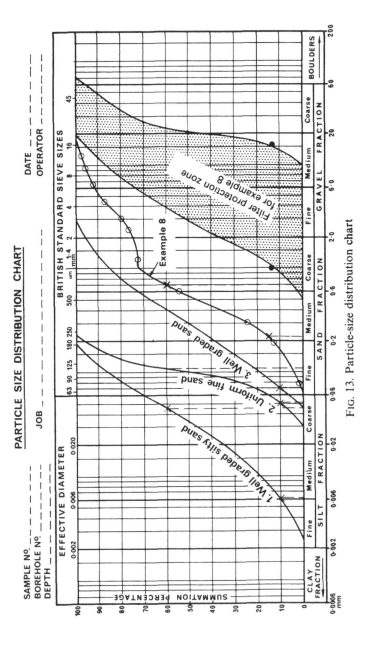

Fig. 13. Particle-size distribution chart

(a) Plot the particle-size distribution curve and describe the soil. Comment on the "flat" part of the curve.

(b) State the effective grain size.

(c) Find Allen Hazen's uniformity coefficient.

(d) Design a filter suitable for protecting this soil.

SOLUTION

Sieve Size (mm)	20	12·5	10	6·3	5·6	2·8	2·0	1·4
Weight passing (g)	147·2	145·5	143·2	134·8	129·1	116·2	112·7	111·6
Percentage passing	100	98·8	97·2	91·5	87·8	79·0	76·5	75·8

Sieve size (μm)	500	355	180	63
Weight passing (g)	81·1	35·8	10·4	3·0
Percentage passing	55·0	24·3	7·0	2·0

For curve plotted from these results *see* Fig. 13.
From Fig. 13:

(a) Description: poorly graded gravelly sand. The "flat" portion of the curve indicates an absence of particles around 2 mm diameter.

(b) D_{10} = Effective size = <u>0·21 mm</u>

(c) $D_{60} = 0.69$ mm ∴ Allen Hazen's uniformity coefficient = $\dfrac{D_{60}}{D_{10}} = 3.3$

(d) For filter design *see* Chapter 4, p. 52.

PLASTICITY

Consistency limits

As moisture is removed from a fine-grained soil it passes through a series of states, i.e. liquid, plastic, semi-solid and solid. The moisture contents of a soil at the points where it passes from one stage to the next are known as *consistency limits*. These limits are defined as:

Liquid limit (LL). The minimum moisture content at which the soil will flow under its own weight.

Plastic limit (P.L). The minimum moisture content at which the soil can be rolled into a thread 3 mm diameter without breaking up.

Shrinkage limit (S.L). The maximum moisture content at which further loss of moisture does not cause a decrease in the volume of the soil.

The range of moisture content over which a soil is plastic is known as the *plasticity index* and is denoted as I_p.

These definitions may be shown diagrammatically (*see* Fig. 14).

Determination of liquid limit

A sample of oven-dried soil, all passing the 0·425 mm sieve, is mixed with distilled water to a stiff consistency, a portion of it placed in

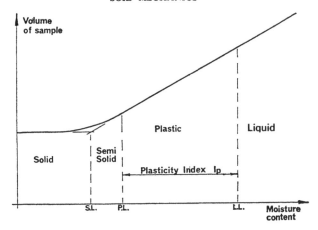

FIG. 14. Consistency-limits graph

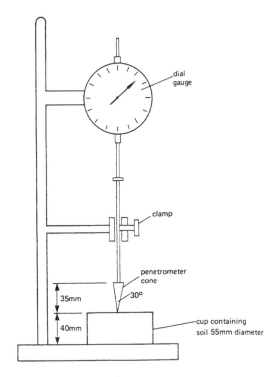

FIG. 15. Standard penetrometer apparatus

the penetrometer cup, (*see* Fig. 15) the soil being struck off level with the top of the cup. The penetrometer cone is then clamped with its tip just touching the soil. The clamp is released and the cone allowed to penetrate the soil for 5 seconds, when the clamp is reapplied. The amount of penetration is read on the dial gauge. This is repeated until two consecutive tests give the same penetration, and this reading is recorded. At this stage the moisture content of the soil in the cup is determined.

The whole procedure is repeated with successive additions of distilled water to the sample, and the relationship between moisture content and penetration plotted on a graph (*see* Fig. 16). The best

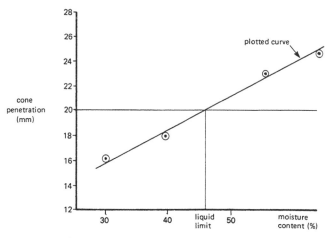

FIG. 16. Plot of penetrometer test results

straight line between these points is drawn and the moisture content corresponding to 20 mm penetration is taken as the liquid limit. This test is described in detail in B.S. 1377 (1975).

Determination of plastic limit
About 20 g of the dried soil, all passing the 0·425 mm sieve, are mixed with distilled water and moulded into a ball. The ball of soil should be rolled by hand on a glass plate with sufficient pressure to form a thread. When the diameter of the resulting thread becomes 3 mm the soil is kneaded together and then rolled out again. The process is continued until the thread crumbles when it is 3 mm diameter, and at this stage the moisture content of the soil is determined. This whole procedure should be carried out twice and the average value of moisture content taken as the plastic limit of the soil. This test is described in detail in B.S. 1377 (1975).

Plasticity chart

The plasticity index of a soil and its liquid limit give one point on a plasticity chart (Fig. 17). Fine-grained soils are subdivided into soils of low, medium and high plasticity as shown, i.e.:

$$\begin{array}{lll}
\text{Low plasticity} & \text{(L)} & \text{L.L.} <35\% \\
\text{Intermediate plasticity} & \text{(I)} & \text{L.L.} 35\%\text{--}50\% \\
\text{High plasticity} & \text{(H)} & \text{L.L.} >50\%
\end{array}$$

The division between inorganic clays and inorganic silts (or organic soils) is by an empirical line (*A* line) the equation of which is I_p 0·73 (L.L. -20). Clays fall above the line and silts below it.

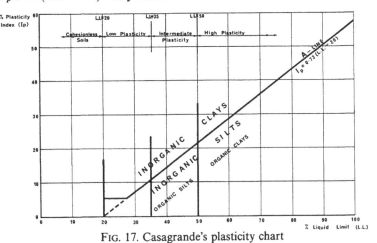

FIG. 17. Casagrande's plasticity chart

CASAGRANDE'S SYSTEM OF SOIL CLASSIFICATION

This system has been developed for roads and airfields. The soil type is designated by two capital letters. (*See* tables 3 and 4)

TABLE 3. CASAGRANDE'S SYSTEM

Main soil type	Prefix		
Coarse-grained soils			
Gravel	G	Well graded	W
Sand	S	Poorly graded	P
		Uniform	U
		Clay binder	C
		Excess fines	F
Fine-grained soils			
Silt	M	Low plasticity	L
Clay	C	Medium plasticity	I
Organic soils	O	High plasticity	H
Peat	Pt		

TABLE 4. EXAMPLES OF CASAGRANDE'S CLASSIFICATION FOR ROADS AND AIRFIELDS

	1	2	3	4	5
	Major divisions	Description and identification	Sub-groups	Casa-grande group-symbol	Applicable classification tests (carried out on disturbed samples)
Coarse-grained soils	Gravel and gravelly soils	Soils with an appreciable fraction between the 80 mm and 2 mm sieves. Generally easily identifiable by visual inspection. A medium to high dry strength indicates that some clay is present. A negligible dry strength indicates the absence of clay	Well-graded gravel-sands with small clay content	GC	Mechanical analysis, liquid and plastic limits on binder
			Uniform gravel with little or no fines	GU	Mechanical analysis
			Poorly-graded gravel-sand mixtures, little or no fines	GP	Mechanical analysis
			Gravel-sand mixtures with excess of fines	GF	Mechanical analysis, liquid and plastic limits on binder of applicable
	Sands and sandy soils	Soils with an appreciable fraction between the 2 mm sieve and the 63 μm sieve. Majority of the particles can be distinguished by eye. Feel gritty when rubbed between the fingers. A medium to high dry strength indicates that some clay is present. A negligible dry strength indicates absence of clay	Well-graded sands and gravelly sands, little or no fines	SW	Mechanical analysis
			Well-graded sand with small clay content	SC	Mechanical analysis, liquid and plastic limits on binder
			Uniform sands, with little or no fines	SU	Mechanical analysis
			Poorly-graded sands, little or no fines	SP	Mechanical analysis
			Sands with excess of fines	SF	Mechanical analysis, liquid and plastic limits on binder if applicable
Finegrained soils Containing little or no coarse-grained material	Fine-grained soils having low plasticity (silts)	Soils with liquid limits less than 35% and generally with less than 20% of clay. Not gritty between the fingers. Cannot be readily rolled into threads when moist. Exhibit dilatancy	Silts (inorganic), rock flour, silty fine sands with slight plasticity	ML	Mechanical analysis, liquid and plastic limits if applicable
			Clayey silts (inorganic)	CL	Liquid and plastic limits
			Organic silts of low plasticity	OL	Liquid and plastic limits from natural conditions after oven drying
	Fine-grained soils having medium plasticity	Soils with liquid limits between 35 and 50% and generally containing between 20 and 40% clay. Can be readily rolled into threads when moist. Do not exhibit dilatancy. Show some shrinkage on drying	Silty clays (inorganic) and sandy clays	MI	Mechanical analysis, liquid and plastic limits if applicable
			Clays (inorganic) of medium plasticity	CI	Liquids and plastic limits
			Organic clays of medium plasticity	OI	Liquid and plastic limits from natural conditions and after oven drying
	Fine-grained soils having high plasticity	Soils with liquid limits greater than 50% and generally with a clay content greater than 40%. Can be readily rolled into threads when moist. Greasy to the touch. Show considerable shrinkage on drying. All highly compressible soils	Highly compressible micaceous or diatomaceous soils	MH	Mechanical analysis, liquid and plastic limits if applicable
			Clays (inorganic) of high plasticity	CH	Liquid and plastic limits
			Organic clays of high plasticity	OH	Liquid and plastic limits from natural conditions and after oven drying

Some examples of these designations can be seen in column 4 of Table 4. This table is based on Casagrande's classification for roads and airfields.

EXAMPLE 9

A liquid limit test carried out on a sample of inorganic soil taken from below the water table gave the following results:

Penetration (mm)	15·5	18·2	21·4	23·6
Moisture content (%)	34·6	40·8	48·2	53·4

A plastic limit test gave a value of 33 per cent.

Determine the liquid limit and plasticity index of this soil and give its classification.

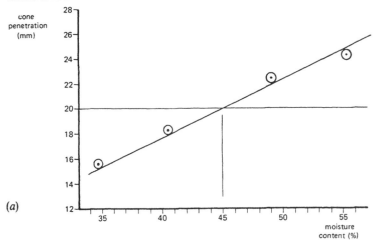

(a)

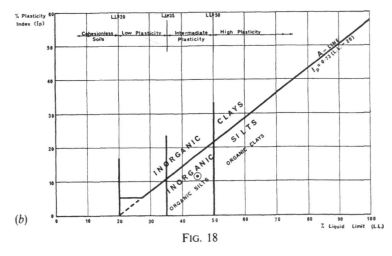

(b)

FIG. 18

SOLUTION
From graph (*see* Fig. 18(*a*)   L.L. = 45
                            P.L. = 33
                            I_p = 12

referring to Fig. 18(*b*), the soil will be classified as an inorganic silt of intermediate plasticity (MI).

QUESTIONS

1. The results of a sieving analysis of a soil were as follows:

Retained on
sieve size (mm) 50   37·5   19   12·5   8   5·9   4·75   2·8
Wt. retained (g) 0   15·5   17   10   11   33   33·5   81

Retained on
sieve size (μm) 2360   1300   400   212   150   100   75
Wt retained (g) 18   31   32·5   9   8   5·5   5

The total weight of the sample was 311g.
Plot the particle-size distribution curve and give a description of the soil

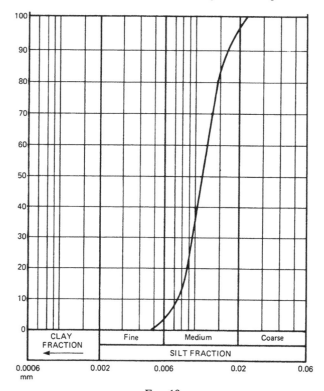

FIG. 19

from this curve. Also find Allan Hazen's effective size and uniformity coefficient.

2. Figure 19 shows a grain-size classification curve for a soil sample. Describe the test carried out to obtain this curve. Describe the soil.

If a suitably prepared sample of this soil was dispersed in a cylinder of water 500 mm deep, estimate how long it would take for all the particles to settle to the bottom of the cylinder.

3. The following results were obtained from a liquid limit test on a fine-grained soil:

Penetration (mm)	15·6	18·2	21·4	23·6
Moisture content (%)	48·6	54·8	62·2	67·4

A plastic limit test gave a value of 22 per cent.

What is the Casagrande classification of this soil?

CHAPTER 4

PERMEABILITY

DARCY'S LAW

Permeability is a measure of the ease with which water flows through rocks and soil. It is of importance to the civil engineer when dealing with seepage under dams, land drainage or groundwater lowering.

The flow of water through soils is assumed to follow Darcy's law:

$$\frac{Q}{t} = kA\frac{H}{l}$$

where Q = quantity of water flowing;
   t = time for quantity Q to flow;
   k = coefficient of permeability for the soil;
   A = area of cross-section through which the water flows;
   H = hydraulic head across soil;
   l = length of flow path through soil.

The ratio H/l is known as the hydraulic gradient and is denoted as i.

The coefficient of permeability k therefore equals.

$$\frac{Q/t}{Ai}$$

and may be defined as the rate of flow per unit area of soil, under unit hydraulic gradient. This coefficient is expressed in mm/s.

Soil type	Values of permeability (mm/s)	Drainage properties
Gravels	$1000 - 10$ ⎫	Good
Sands	$10 - 10^{-2}$ ⎭	
Silts (and fissured clay)	$10^{-2} - 10^{-5}$	Poor
Clays	$10^{-5} \longrightarrow$	Impervious

DETERMINATION OF THE COEFFICIENT OF PERMEABILITY

Coarse-grained soils

Constant-head permeameter. Water under a constant head of pressure is allowed to percolate through a sample contained in a cylinder of cross-sectional area A. The quantity of water Q passing through the sample in time t is collected in a measuring cylinder. Manometers tapped into the side of the sample cylinder give the loss of head H over a length of sample l and hence the hydraulic gradient i. From Darcy's law:

$$\text{Coefficient of permeability} = \frac{Q/t}{Ai}$$

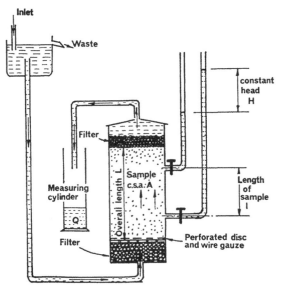

FIG. 20. Constant-head permeameter

This apparatus is known as a *constant-head permeameter* and is shown in Fig. 20. The water may be arranged to flow up the sample as shown, but some permeameters permit downward flow, the same principles applying. A sand filter is incorporated above and below the sample to help prevent it washing away.

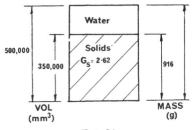

FIG. 21

EXAMPLE 10

A constant-head permeameter test has been run on a sand sample 250 mm in length and 2000 mm^3 in area. With a head loss of 500 mm the discharge was found to be 260 ml in 130 seconds. Determine the coefficient of permeability of the soil.

If the specific gravity of the grains was 2·62 and the dry weight of the sand 916 g, find the void ratio of the sample.

SOLUTION

$$k = \frac{260 \times 1000/130}{2000 \times 500/250} = \underline{\underline{0\cdot5 \text{ mm/s}}}$$

From Fig. 21

$$V_s = \frac{(916/1000) \times 1000^3}{2\cdot62 \times 1000} = 350\,000 \text{ mm}^3$$

$$V_v = 250 \times 2000 - 350\,000 = 150\,000 \text{ mm}^3$$

$$e = \frac{V_v}{V_s} = \frac{150\,000}{350\,000} = \underline{\underline{0\cdot428}}$$

In situ value of permeability. It may be seen from Example 10 that if the dry weight of the sample in the permeameter and the specific gravity of the grains are known the coefficient of permeability for varying values of void ratio may be determined.

Fill the permeameter loosely for the first determination and then tap down the sample to decrease the voids for succeeding tests. It should be noted that the length of sample is taken as the overall length L for determining volume.

If a graph of e against $\log_{10} k$ is plotted, a straight line is frequently obtained (*see* Fig. 22).

If the site value of void ratio e is known the value of k corresponding to this *in situ* value of e can be read off the graph.

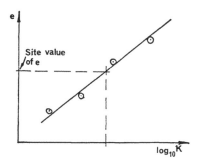

FIG. 22. *In situ* value of permeability; $e \longrightarrow \log k$ graph

Fine-grained soils

Variable-head permeameter
Water flows through fine-grained soils at a much slower rate than through coarse material; consequently, it is not possible to obtain a measurable amount of water within a reasonable time. In this case a *variable-head permeameter* (*see* Fig. 23) is used.

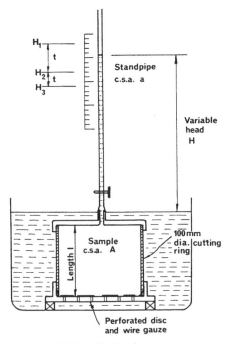

FIG. 23. Variable-head permeameter

When the stopcock is opened, water will pass through the sample and the level in the standpipe will fall. Once steady conditions have been obtained, two readings of H are taken, H_1 and H_2 at a time interval t.

Now during any increment of time dt the variation in head is $-dH$, and hence the quantity of water flowing through the sample in time dt, $Q = -a\,dH$. From Darcy's law:

$$\frac{-adH}{dt} = kA\frac{H}{l}$$

$$-a\,dH = kA\frac{H}{l}dt$$

or
$$dt = -\frac{al}{Ak}\frac{dH}{H}$$

integrating between limits 0 to t and H_1 to H_2

$$-t = \frac{-al}{Ak}\ \log_e\frac{H_1}{H_2}$$

or
$$k = \frac{a}{A} \times \frac{l}{t} \times 2\cdot3\ \log_{10}\frac{H_1}{H_2}$$

In this equation all the terms may readily be found and hence k calculated.

Three readings of H should be taken H_1, H_2 and H_3 such that the time for the head to drop from H_1 to H_2 is the same as the time for the head to drop from H_2 to H_3. Since in the permeability equation derived k, a, A and l are constant and t has also been made the same, then:

$$\log_{10}\frac{H_1}{H_2} = \log_{10}\frac{H_2}{H_3}$$

or
$$\frac{H_1}{H_2} = \frac{H_2}{H_3}$$

$$\therefore\ H_2 = \sqrt{H_1\ H_3}$$

This should be checked when carrying out the test to ensure that steady conditions have been obtained.

The permeability of clay soils cannot be found by direct laboratory testing, but an estimate may be made indirectly from consolidation test results (*see* Chapter 5).

Multi-layer permeability
In natural conditions soil is rarely, if ever, homogeneous. Even in apparently homogeneous soils, stratification will occur giving thin layers of varying permeability. On the larger geological scale, the strata may vary widely from a relatively impervious clay to a permeable sand within a small depth. Similarly, man-made filters may be graded from coarse to fine material in layers.

These variations will have a marked effect on the overall permeability, with the average value in the direction of stratification being quite different from the value at right-angles to it.

In a series of strata, thickness H_1, H_2, H_3 etc. with permeabilities k_1, k_2, k_3 etc. the rate of flow per unit area along each stratum will vary but the hydraulic gradient will be constant. The average permeability in this direction can be shown to be equal to k_H where:

$$k_H = \frac{k_1H_1 + k_2H_2 + k_3H_3 + \cdots\cdots k_nH_n}{H_1 + H_2 + H_3 + \cdots\cdots H_n}$$

With flow at right-angles to the strata, the hydraulic gradient will vary in each stratum, but the rate of flow per unit area must be constant. The average permeability at right-angles to the strata can be shown to be equal to k_v where:

$$k_V = \frac{H_1 + H_2 + H_3 + \cdots\cdots\cdots H_n}{H_1/k_1 + H_2/k_2 + H_3/k_3 + \cdots\cdots H_n/k_n}$$

From these two equations it can be proved that $k_H/k_V > 1$, i.e. that the permeability in the direction of the strata k_H is always greater than the permeability at right-angles to the strata k_V.

As soil samples for laboratory testing are frequently taken at right-angles to the strata, it can be seen that laboratory tests can give a low value of the actual permeability on the site.

EXAMPLE 11

In a falling head permeameter test on a silty clay sample, the following results were obtained: sample length 120 mm; sample diameter 80 mm; initial head 1100 mm; final head 420 mm; time for fall in head 6 minutes; standpipe diameter 4 mm.

Determine from first principles the coefficient of permeability of the soil.

On close investigation of the sample it was found to be in 3 layers 20 mm, 60 mm and 40 mm thick, each of permeabilities 3×10^{-3} mm/s, 5×10^{-4} mm/s and 17×10^{-4} mm/s respectively.

Check the average permeability through the sample in the laboratory test and estimate the permeability of this sample in a direction at right-angles to sampling.

Find the ratio k_H/k_V and comment on the result.

SOLUTION

The derivation of the expression $k = \dfrac{a}{A} \times \dfrac{l}{t} \times 2 \cdot 3 \log_{10} \dfrac{H_1}{H_2}$ has been given in the previous section. Substituting the values given into this expression,

$$k = \frac{\pi 4^2/4}{\pi 80^2/4} \times \frac{120}{\cdot 360} \times 2 \cdot 3 \log_{10} \frac{1110}{420}$$

$$= \underline{\underline{8 \times 10^{-4} \text{ mm/s}}}$$

For three layer case, permeability of laboratory sample,

$$k_V = \frac{20 + 60 + 40}{20/(3 \times 10^{-3}) + 60/(5 \times 10^{-4}) + 40/(17 \times 10^{-4})}$$

$$= \underline{\underline{7 \cdot 99 \times 10^{-4} \text{ mm/s}}}$$

which is as given by the original test.

Permeability in a direction at right-angles to sampling,

$$k_H = \frac{3 \times 10^{-3} \times 20 + 5 \times 10^{-4} \times 60 + 17 \times 10^{-4} \times 40}{20 + 60 + 40}$$

$$= 1 \cdot 3 \times 10^{-3} \text{ mm/s}$$

$$\frac{k_H}{k_V} = \underline{\underline{1 \cdot 6}}$$

It can be seen that if the flow of water on site is along the strata, the laboratory results give an underestimate of the flow that will occur.

Accuracy of permeability measurements

It should always be remembered that the measurement of permeability of a soil can never be accurate. Firstly there is the enormous range of values covered, from 1000 mm/s for a coarse gravel down to say 0·00001 mm/s for a fine silt, which limits the degree of accuracy normally obtained in calculations.

The wide variation of soil on a site will cause variations in permeability, depending on the direction of flow, as has just been shown. As an exercise, the student should consider a 5 m layer of silt with a laboratory measured permeability of 3×10^{-5} mm/s. This layer of silt will not be homogeneous but may be in 1 mm thick layers, each with a slightly different value of permeability. If say *only 3 layers* each 1 mm thick have a permeability of 8×10^{-3} the value k_H/k_V will be approximately 9!

Finally, there is the serious problem of sampling. Only very small samples are used to estimate the permeability of a large site and it is necessary to test a large number of representative samples. Moreover samples will also be disturbed. In the case of the constant head permeameter using coarse grained soils, the disturbance will be acute and this test is only really suitable for man-made "soils" such as filter. In the falling head permeameter, sample disturbance also occurs, and it is difficult to be sure that the effects of seepage between the soil and its container are negligible.

Generally therefore, site tests give a far more satisfactory result than laboratory tests.

DETERMINATION OF PERMEABILITY ON SITE

Borehole techniques

An estimate of the permeability of a soil may be made using the boreholes driven during the site investigation. There are many empirical techniques for determining permeability in this way and, given the general inaccuracy of determining permeability, these are reasonably satisfactory methods.

Generally, if the stratum being tested is above the water table, water is pumped into the bore-hole and the rate of flow to maintain a constant head is measured. If the stratum is below the water table, either pumping in or pumping out tests may be used, in conjunction with a casing to the bore-hole extending into the permeable stratum.

The theoretical approach to these determinations is beyond the scope of this volume, but the U.S. Bureau of Reclamation use the following expression for cased bore-holes:

$$k = \frac{q}{5 \cdot 5 \, rh}$$

where k = coefficient of permeability
         q = rate of flow of water into bore-hole to maintain constant head above the water table
         r = radius of the casing
         h = head of water maintained above the water table

Such methods however, rely on the experience of the engineer rather than mathematical accuracy to obtain a reasonable result.

Well-point techniques

If a well-point method of ground-water lowering is used it is possible to determine the coefficient of permeability in the field.

When water is pumped from a well point the water is lowered adjacent to the point, giving a cone of depression. This cone of depression will form even in relatively impervious soils after sufficient time has elapsed.

Consider the horizontal flow of water through a thin element of the soil at distance r from the well point where the head of water above an impervious layer is Z (*see* Fig. 24).

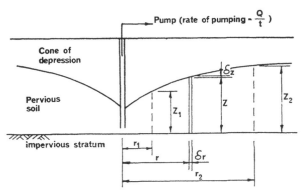

FIG. 24. Ground-water lowering: well point (I)

The rate of flow, Q/t, is the rate at which water is being pumped and may readily be determined.

The surface area of the soil element at distance r from the well point is equal to the surface area of a cylinder radius r height z, or $A = 2\pi rz$.

The hydraulic gradient at the point is the rate of change of head per unit length or, $i = dz/dr$. From Darcy's law:

$$Q/t = k \, Ai$$

$$Q/t = k \times 2\pi rz \times \frac{dz}{dr}$$

or

$$\frac{1}{r}dr = k \times \frac{2\pi}{Q/t} \times z \, dz$$

Integrating between limits r_2 to r_1 and z_2 to z_1:

$$\log_e\frac{r_2}{r_1} = k \times \frac{2\pi}{Q/t} \times \frac{(z_2{}^2 - z_1{}^2)}{2}$$

$$k = \frac{Q/t \log_e\dfrac{r_2}{r_1}}{\pi(z_2{}^2 - z_1{}^2)}$$

$$= 2 \cdot 3\frac{Q/t \log_{10}\dfrac{r_2}{r_1}}{\pi(z_2{}^2 - z_1{}^2)}$$

Hence, by measuring the height of the ground water at two points distances r_1 and r_2 from the well point the coefficient of permeability may be determined.

Considerable practical experience is required before a reliable result by this method may be obtained, since the soil is unlikely to be homogeneous, nor is any impervious layer present likely to be horizontal. Pumping should be allowed to continue until conditions are settled before any measurements are taken, and the observation wells must not be too close to the well point, where soil will be disturbed and the drop in head too rapid.

The student should now develop the equation for determining the coefficient of permeability for a permeable stratum thickness d and overlain by a relatively impervious stratum (see Fig. 25).

$$k = \frac{2 \cdot 3 \, Q/t \log_{10}\dfrac{r_2}{r_1}}{2\pi d(z_2 - z_1)}$$

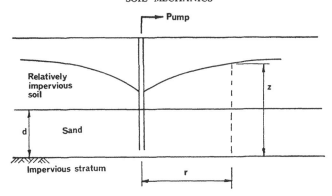

FIG. 25. Ground-water lowering: well point (II)

EXAMPLE 12

A stratum of sandy soil overlies a horizontal bed of impermeable material, the surface of which is also horizontal. In order to determine the *in situ* permeability of the soil, a test well was driven to the bottom of the stratum (*see* Fig. 26). Two observation bore holes were made at distances 12 and 24 m respectively from the test well. Water was pumped from the test well at the rate of 180 litres/min until the water level became steady. The heights of the water in the two bore holes were then found to be 4·2 and 6·3 m above the impermeable bed. Find the value, expressed in mm per second, of the coefficient of permeability of the sandy soil, deriving any formula used.

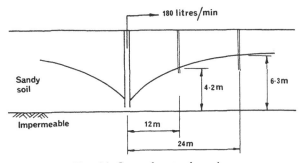

FIG. 26. Ground-water lowering

SOLUTION

Derivation of formula:

$$k = \frac{2 \cdot 3 \, Q/t \, \log\dfrac{r_2}{r_1}}{\pi(z_2{}^2 - z_1{}^2)}$$

$$\text{Coefficient of permeability} = \frac{2 \cdot 3 \times (180 \times 1000^2)/60 \times \log\frac{24}{12}}{\pi(6 \cdot 3^2 - 4 \cdot 2^2) \times 1000^2}$$

$$= \underline{\underline{0 \cdot 03 \text{ mm/s}}}$$

CRITICAL HYDRAULIC GRADIENT

The critical hydraulic gradient is the hydraulic gradient at which the soil becomes unstable, i.e. when the intergranular pressure (effective stress) becomes zero.

Consider a sample of soil, length d, with water flowing upwards owing to head h as shown in Fig. 27.

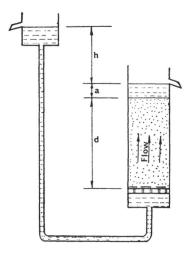

FIG. 27. Critical hydraulic gradient

Hydraulic gradient $= h/d$

Total downward load at base of sample
$$= d\gamma_{sat} + a\gamma_w$$

Neutral stress $= (h + a + d)\gamma_w$

Effective stress (or intergranular pressure)
$$= d\gamma_{sat} + a\gamma_w - (h + a + d)\gamma_w$$
$$= d(\gamma_{sat} - \gamma_w) - h\gamma_w$$
$$\sigma' = d\gamma' - h\gamma_w$$

$d\gamma'$ is the submerged weight of soil and must be greater than $h\gamma_w$ for there to be any intergranular pressure. $h\gamma_w$ is known as the *seepage pressure*.

If the head h is increased until $d\gamma' = h\gamma_w$, then $\sigma' = 0$ and the soil will become unstable. In this condition the hydraulic gradient $h/d = \gamma'/\gamma_w$ and is known as the critical hydraulic gradient i_c. Also:

Critical hydraulic gradient $i_c = \gamma'/\gamma_w$

$$= \frac{\gamma_{sat} - \gamma_w}{\gamma_w}$$

$$= \frac{\left(\dfrac{G_s + e}{1 + e}\right)\gamma_w - \gamma_w}{\gamma_w}$$

(*see* Example 4, Chapter 1)

or $$i_c = \frac{G_s - 1}{1 + e}$$

Quicksand

A soil under critical hydraulic gradient will be unstable and is said to be in a "quick" condition. By this definition any granular soil may be a "quick sand", but soils with high permeability (i.e. gravels and coarse sands) require large quantities of water to maintain a critical hydraulic gradient Quicksand conditions are therefore usually confined to fine-grained sands.

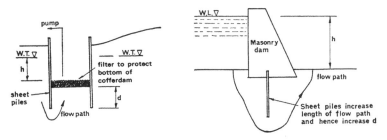

FIG. 28. Examples of possible quick conditions in engineering works

Filter design

If upward flow of water is taking place, to help prevent quick conditions occurring a load should be placed on the surface of the soil, thus increasing the effective pressure. This load should consist of a coarser material (filter) than the soil it is stabilising and is subject to the following limiting conditions:

1. It must be sufficiently coarse to become readily saturated, thus avoiding seepage forces being set up.
2. It must be fine enough to prevent passage through its pores of particles from the soil it is stabilising.

Item 1 is satisfied if: D_{15} for the filter material is greater than four to five times D_{15} for the soil it is protecting.

Item 2 is satisfied if: D_{15} for the filter material is less than four to five times D_{85} for the soil it is protecting.

Using the figures from Example 8, p. 33, from the particle-size distribution curve for this soil:

$D_{15} = 0.22$ mm $0.22 \times 4 = 0.88$ mm $0.22 \times 5 = 1.1$ mm
$D_{85} = 4.3$ mm $4.3 \times 4 = 17.2$ mm $4.3 \times 5 = 21.5$ mm

therefore, for the filter material, D_{15} should lie between the limits of 1·1 and 17·2 mm.

Curves drawn approximately parallel to the curve plotted for the soil but passing through $D_{15} = 1.1$ and $D_{15} = 17.2$ are shown in Fig. 13. The filter material should have a curve that falls within these limits.

Filters designed in this way are used in many circumstances. As shown (*see* Fig. 28), they stabilise a sand liable to quick conditions, and this may occur at the bottom of an excavation, alongside a river wall, downstream of a dam or in any conditions where water is flowing upward through a soil.

If the filter material needs to be very fine, to prevent the particles from the protected soil passing, it is better to lay only a thin layer, and then protect this filter material with a coarser soil. Sometimes a filter may be constructed in several layers, each about 300 mm thick, and each layer designed to protect the layer below it. This is known as a *reversed* or *graded* filter.

SEEPAGE THROUGH SOIL

In water-retaining dams, unless the foundations continue down to impervious rock, a steady flow of water is set up under the structure owing to the difference in head. This may lead to an undesirable amount of leakage and, with upward flow of water on the downstream side, dangerous quicksand conditions may occur, with possible subsequent failure of the dam. This seepage can be studied by the use of flow nets.

FLOW NETS

A flow net is a pictorial representation, drawn to scale, of the paths taken by water in passing through a material. It is made up of flow lines and equipotential lines (*see* Fig. 29).

Flow lines. These represent the path of flow through a soil. There are an infinite number of flow lines, the paths of which never cross (laminar flow). Each impervious boundary may be taken as a flow line, and a few selected intermediate paths are drawn, each line being approximately parallel to the last.

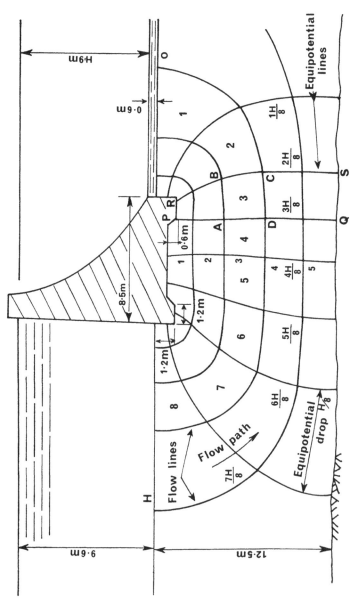

FIG. 29. Flow net (scale 1:250)

Equipotential lines. Water will only flow where there is a pressure head, say H, and this is dissipated as water flows through the soil. On each flow line, therefore, there will be a point where the pressure head has dissipated to (say) $\frac{7}{8}H$. All these points where the pressure head is equal may be joined by an *equipotential line.* There are an infinite number of these equipotential lines, but for graphical construction only a few are drawn. At the point where water flows into and out of the soil the ground levels will be equipotential lines.

Water will flow along the path in which H/l, the hydraulic gradient, is a maximum. Since the head dissipated between any two adjacent equipotential lines is constant ($\frac{1}{8}H$ in Fig. 29), for maximum hydraulic gradient the minimum value of l must be used. This minimum value of l will be when flow lines are at right angles to equipotential lines. Hence a flow net will be composed of a series of approximate rectangles, which are called *fields.*

When drawing a flow net it is advisable to choose flow lines and equipotential lines to give approximately square fields, since these are easier to recognise. There will always be a few fields at the boundaries which do not approximate to squares, and these are known as *singular* fields. Further subdivision should yield truer squares.

Construction of flow nets

Before any further examples are attempted the following points should be noted concerning the construction of flow nets:

1. *Flow lines* should be drawn with each one approximately parallel to the last. Flow lines will *never* cross one another.
2. *Equipotential lines* are drawn such that they cross the flow lines at right angles. It is easier to select the equipotential lines which form approximately square fields.

Some examples of flow nets are shown in Figs. 29, 31 and 32. The student should practise drawing flow nets which conform to the conditions given above, giving special consideration to the boundary fields.

Use of flow nets

Flow nets may be used to determine the rate of loss of water from a reservoir, or the magnitude of the seepage pressure and hence the possibility of instability of the soil.

Loss of water due to seepage

Let N_f = Number of flow paths
N_e = Number of equipotential drops

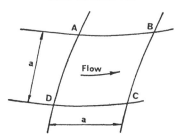

FIG. 30. Flow net field. (Taken from Fig. 29.)

Now consider one square, side a, and over a length of the dam equal to unity (*see* Fig. 30).

Let the loss of head from AD to $BC = dh$

$$\text{where } dh = \frac{H}{N_e} \qquad (\tfrac{1}{8}H \text{ in Fig. 29})$$

From Darcy's law:

$$\frac{Q}{t} = kA\frac{H}{l}$$

or rate of flow from AD to BC over unit width $= k \times 1a \times \dfrac{dh}{a}$

$$= k\,dh$$

∴   *rate of flow from PQ to RS* (Fig. 29) over unit width $= k\,dh\,N_f$

but $$dh = \frac{H}{N_e}$$

$$\frac{Q}{t} = k\,H\frac{N_f}{N_e}$$

EXAMPLE 13

(*a*) If the proposed dam shown in Fig. 29 is 90 m long and the coefficient of permeability of the soil is 0·0013 mm/s, find the quantity of water that will be lost per day by seepage.

(*b*) To decrease this loss, sheet piles were driven in at the toe of the dam to a depth of 5·8 m, as shown in Fig. 31, and an impervious apron constructed at the heel 6 m wide. What will be the loss of water per day when these measures are taken?

SOLUTION
See Fig. 29.

(*a*) $$\frac{Q}{24 \times 60 \times 60} = \frac{0·0013}{1000} \times 9 \times \tfrac{5}{8} \times 90$$

$$Q = 57 \text{ cubic metres per day}$$

∴   Rate of loss of water $= 57\,000$ litres per day

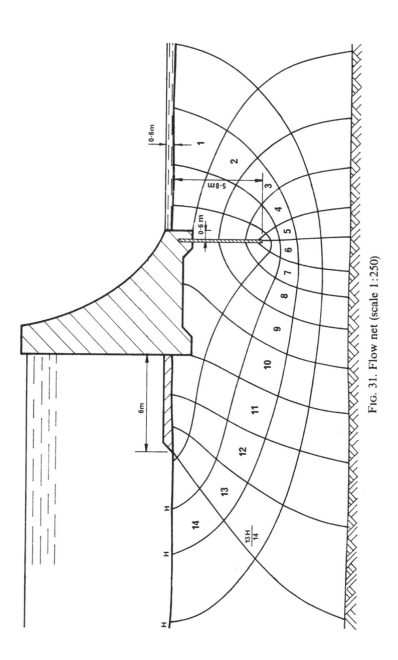

FIG. 31. Flow net (scale 1:250)

(*b*) The effect of these measures is to considerably increase the length of path of flow, and the new flow net will be as shown in Fig. 31.

Since all the terms in the permeability equation will be as before, except N_e, which is now 14, the new value for rate of loss of water

$$= 57\,000 \times \tfrac{8}{14}$$
$$= 32\,000 \text{ litres/day}$$

Instability due to seepage pressure
Referring to Fig.30:

Seepage pressure from AD to $BC = dh\,\gamma_w$

or   seepage force from AD to $BC = dh\,\gamma_w \times 1a$

$$= \frac{dh}{a}\gamma_w \times 1a^2$$

but dh/a is the hydraulic gradient; $1a^2$ is the volume of soil; therefore the seepage force per unit volume $= i\gamma_w$

When the water is flowing downwards the seepage pressure causes an increase in intergranular pressure. When the water flows upwards, however, the intergranular pressure is reduced, and therefore there is a tendency towards unstable conditions on the downstream side of a dam.

EXAMPLE 14
A sheet-pile wall is driven to a depth of 6 m into permeable soil which extends to a depth of 13·5 m below ground level. Below this there is an impermeable stratum. There is a depth of water of 4·5 m on one side of the sheet-pile wall. Make a neat sketch of the flow net and determine the approximate seepage under the sheet-pile wall in litres per day, taking the permeability of the soil as 6×10^{-3} mm/s.

Explain the term *piping* and show how the flow net can be used to determine whether this condition is likely to occur in front of the sheet piling.

(Assume soil density of 1900 kg/m³.)

SOLUTION
Referring to Fig. 32:

from the flow net

$$N_f = 5 \qquad\qquad N_e = 10$$
$$\therefore \quad Q/t = 6 \times 10^{-3} \times 4{\cdot}5 \times \tfrac{5}{10}$$
$$= 0{\cdot}0135 \text{ litres/second/metre length}$$
$$= 0{\cdot}0135 \times 60 \times 60 \times 24$$
$$= 1166 \text{ litres/day/metre length}$$

For explanation of piping, or quicksand, *see* p. 52.

It has been found in practice that piping is liable to take place in front of sheet piles for a distance of about half the depth of penetration, i.e. for this

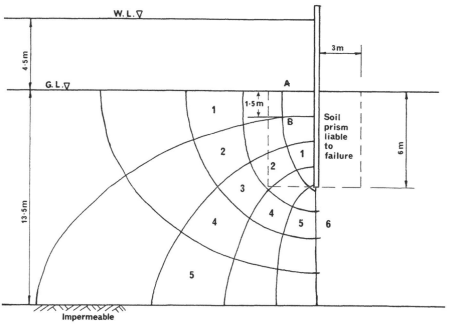

FIG. 32. Flow net

example a prism 6 m deep × 3 m wide × 1 m thick is possibly subject to piping.

Consider the flow line in the area most liable to failure. (*AB* is used in Fig. 32 since the flow net is symmetrical.)

Hydraulic gradient $\qquad i_{AB} = \dfrac{h_{AB}}{l_{AB}}$

where $\qquad\qquad h_{AB} = \dfrac{4\cdot5}{10} \qquad\qquad l_{AB} = 1\cdot5 \text{ m}$

$$\therefore \quad i_{AB} = \frac{4\cdot5}{10 \times 1\cdot5} = 0\cdot3$$

Critical hydraulic gradient $i_c = \dfrac{\gamma'}{\gamma_w} = \dfrac{1900 - 1000}{1000} = 0\cdot9$

$$\therefore \quad \text{factor of safety against piping} = \frac{0\cdot9}{0\cdot3} = 3\cdot0$$

which would be quite satisfactory.

QUESTIONS

1. The following results were obtained from a falling head permeameter test on a sandy silt: sample length 140 mm; sample diameter 70 mm; initial

head 1400 mm; final head 220 mm; time for fall in head 1 minute 20 seconds; standpipe diameter 6 mm.

A constant head test is carried out on the same soil compacted to the same void ratio. Calculate the quantity of water flowing through the sample in 10 minutes if the head of water over a 100 mm length of sample is 30 mm. The internal diameter of the permeameter is 90 mm.

2. Discuss the difficulties of estimating the site value for the coefficient of permeability of a soil in laboratory tests.

A graded filter is constructed of 4 layers of soil. The layers are 400 mm, 300 mm, 140 mm and 60 mm thick and compacted to give permeabilities of 2×10^{-2} mm/s, 4×10^{-1} mm/s, 8×10^{-1} mm/s and 1 mm/s respectively. Calculate the average coefficient of permeability in directions parallel to and at right angles to the layers.

3. A well-point lowering scheme is carried out on a site and after steady conditions have been obtained the readings in observation wells are as shown in Fig. 33. If the rate of pumping is 100 litres/minute, estimate the coefficient of permeability of the sand stratum.

Draw a flow net for this well-point system and use it to check the value of permeability calculated above.

Note: A plan view radial flow net is required for this calculation.

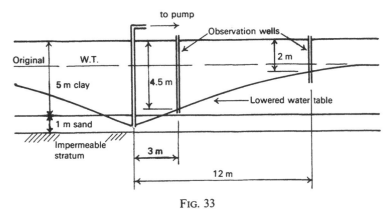

FIG. 33

4. A falling-head permeameter test carried out on a sample of the permeable soil shown in Fig. 34 produced the following results: length of sample 101·6 mm; diameter of sample 73 mm; initial head 1352 mm; final head 352 mm; time interval 147 sec; diameter of standpipe 5 mm. Construct the flow net for the conditions shown in the figure and estimate the seepage loss in litres/day/10 metre length of the coffer dam.

5. Determine by means of a neat flow diagram the seepage loss in litres/day/metre run of wall beneath the sheet piling shown in Fig. 35. Permeability of the soil is 0·002 mm/s.

6. What do you understand by the term "flow net"?

A 12 m-thick deposit of cohesionless soil of permeability 0·035 mm/s has

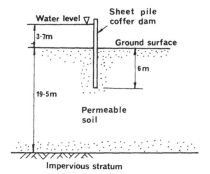

FIG. 34

a level surface and overlies an impermeable layer. A long row of sheet piles is driven 6 m into the soil. The wall extends above the surface of the soil, and impounds a depth of 3·7 m of water on one side: the water level on the other side is maintained at ground level. Sketch the flow net and determine the seepage quantity per metre run of wall, deriving any formula used.

What is the value of the pore water pressure at a point near the toe of the wall?

How would you investigate the factor of safety against piping in this problem?

7. Draw to a scale of 1:100 an accurate flow net to represent flow under the sheet pile wall in Fig. 36. Assume that there is no flow through the sheet piling, and that the sand is of uniform permeability. If the difference in head between upstream and downstream water levels is H metres, plot a graph showing the seepage pressure varies along the concrete apron.

8. Fig. 37 shows a long coffer dam consisting of two rows of sheet piles in sand. Estimate, by means of a neat flow diagram, the rate in mm/hour at which the water level due to seepage will rise in the coffer dam immediately after pumping the coffer dam dry. Take the coefficient of permeability to be 0·1 mm/s.

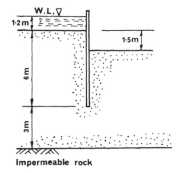

FIG. 35

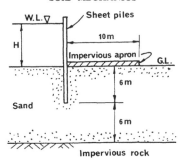

FIG. 36

Investigate the possibility of a quicksand condition forming if the saturated density is 1920 kg/m^3.

9. A reinforced concrete wall is to be built to retain water in a reservoir located on 12 m depth of permeable soil which overlies an impermeable stratum. The base of the wall is of rectangular cross-section 5 m wide by 1 m deep and the top of the base is at ground level. It is estimated that the seepage loss will be 80 000 litres per day. Sketch a flow net to illustrate seepage in this condition.

In order to reduce this seepage loss it is decided to extend the width of the base by 2 m and drive a continuous sheet pile wall below the proposed extension as shown in Fig. 38. Draw a flow net for this condition and estimate the reduction in seepage loss. What will be the percentage change in the factor of safety against quicksand conditions?

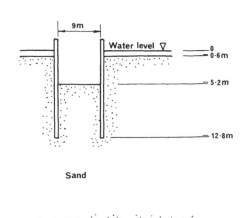

FIG. 37

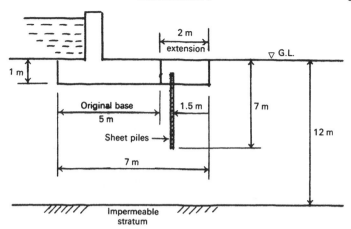

FIG. 38

CHAPTER 5

CONSOLIDATION

SETTLEMENT

Settlement is the most common reason for failure of foundations and it is therefore of great importance to understand the mechanics of settlement. The elastic deformation of the ground when loaded will occur immediately the load is applied, and can therefore be relatively easily corrected. The long-term consolidation of clay soils, however, may take many years to complete, with consequent damage to the structure appearing long after completion of the contract. In coarse-grained soils consolidation occurs rapidly and is therefore a different problem. There may also be further settlement when consolidation nears completion due to "creep" in the soil mass, this is known as secondary settlement.

The long-term consolidation of clay soils is probably the most troublesome type of settlement and is the main consideration in this chapter.

OEDOMETER TEST

In fine-grained soils, much useful information can be obtained from a laboratory oedometer test.

An undisturbed sample of soil is retained in a 75-mm-diameter cutting ring (*see* Fig. 39) with drainage at both faces. The soil is kept saturated throughout the test.

1. A load p_1 is applied to the sample and the change in thickness (compression) of the sample read at suitable intervals, up to 24 hours.

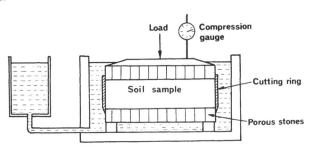

Load Compression gauge

Soil sample

Cutting ring

Porous stones

FIG. 39. Oedometer.

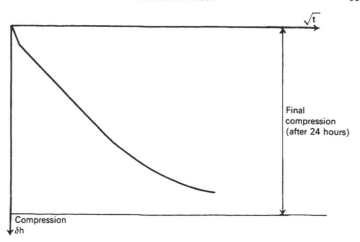

FIG. 40. Square root of Time⟶Compression graph

2. A graph of square root of time against compression is plotted and should be as shown in Fig. 40.

3. The load is now increased to p_2 and another graph of compression against square root of time plotted for 24 hours.

4. This procedure is repeated with loads p_3, p_4, etc. until a sufficient range of load has been covered. Typical values of p_1, p_2, p_3, etc. are 50, 100, 200, 400, 800 kN/m² , each applied for 24 hours. It is usually only necessary to plot the first few hours of each test on the graph, but the 24-hour reading of compression *must* be taken.

5. When sufficient range of loading has been covered the load is removed and the sample allowed to take in water and expand for 24 hours. A graph is not normally plotted for this expansion period, but the final thickness of the sample h_f *must* be recorded.

6. When expansion is complete the final moisture content m_f of the sample *must* be determined.

As the sample is saturated the final void ratio of the soil e_f may be determined.

$$e_f = m_f G_s$$

The oedometer test gives a guide both to the amount of consolidation which will occur on site and the rate at which it proceeds. It is described in detail in B.S. 1377 (1975).

THEORY OF CONSOLIDATION

The full theory of consolidation is beyond the scope of this book, but it is of value to have an idea of the mechanics of consolidation and the assumptions on which the theory is based. It is also necessary to know certain constants and definitions which derive from the theory.

Since the clay stratum in which consolidation occurs will frequently be below the water table it is assumed the clay is saturated. Fig. 41(a) shows a clay stratum beneath the water table and the effective stress equation at any given level A–A in this stratum will be

$$\sigma = \sigma' + u$$

where σ is the total vertical pressure, σ' is the effective stress and u is the neutral stress or water pressure at level A–A. The stress diagrams for an example of this condition are shown in Fig. 41(a).

When a load p is applied to a saturated soil, in the first instance, the whole of this load is carried by the water. The effective stress equation then becomes

$$(\sigma + p) = \sigma' + (u + p)$$

and the pressure diagram within the clay stratum are shown for this condition in Fig. 41(b).

In the condition shown in Fig. 41(b) the soil water is subjected to pressure head p and this pressure head will begin to dissipate. In the overlying sand this will happen rapidly owing to the high permeability of the sand. Due to the low permeability of the clay however drainage will be most rapid where a clay to sand boundary occurs and may be considered negligible within the clay itself. After a period of time t therefore some of the excess water pressure will have dissipated at the boundaries and this pressure will have transferred to the soil grains as effective stress. The equation of effective stress after time t therefore may be expressed as

$$(\sigma + p) = (\sigma' + \delta p) + (u - \delta p)$$

and the pressure diagrams are shown in Fig. 41(c). It is this increase in effective stress (intergranular pressure) which causes the soil to consolidate.

Consolidation is considered complete when the whole of the excess pressure is transferred to the soil grains and the effective stress equation becomes $(\sigma + p) = (\sigma' + p) + u$. All the excess neutral stress is now dissipated and the pressure diagram is shown in Fig. 41(d).

Assumptions of theory of consolidation
This concept makes certain assumptions such as:

1. the clay is saturated,
2. drainage can only occur in the vertical direction,

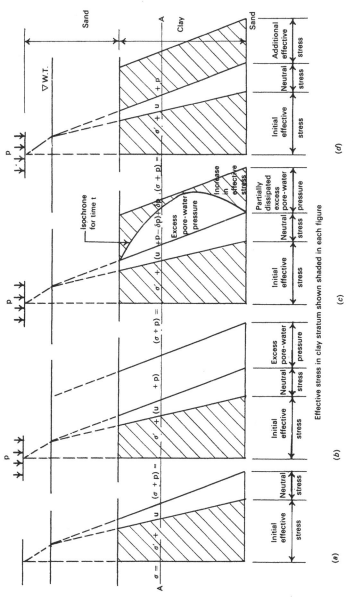

Effective stress in clay stratum shown shaded in each figure

FIG. 41. Variation of pressure during consolidation

3. Darcy's law is valid for fine grained soils,
4. solid particles are incompressible,
5. total stresses remain constant on any horizontal plane during consolidation,
6. the clay does not expand laterally.

It is these assumptions which make it impossible to predict consolidation accurately, and they should be borne in mind when estimating settlement.

Constants and definitions arising from the theory of consolidation
Coefficient of compressibility m_v is defined as the decrease in unit volume per unit increase of pressure, i.e.:

$$m_v = \frac{-\delta V}{V_0} \bigg/ \delta p$$
$$= \frac{-1}{V_0} \cdot \frac{\delta V}{\delta p}$$

where V_0 = the original volume and δp the change in pressure.

This constant is concerned with volume change and therefore is used to determine the total *amount* of consolidation settlement. It will vary depending on the applied range of pressure.

Coefficient of consolidation c_v is a constant derived from the theory where:

$$c_v = \frac{k}{m_v \gamma_w}$$

where k is the permeability of the soil, m_v the coefficient of compressibility and γ_w the density of water ($9 \cdot 8$ kN/m^2).

This constant contains the value of permeability, which determines the *time* required for consolidation to occur. Since both k and m_v will reduce during consolidation it is assumed that they reduce proportionately and therefore k/m_v is constant. This means that c_v has a constant value, but this assumption should be treated with extreme caution.

Both of these constants m_v and c_v are determined from the oedometer.

Degree of consolidation U_v is defined as the percentage of the total final settlement that has occurred after time t.

$$U_v = \frac{\text{Settlement after time } t}{\text{Total final settlement}} \times 100$$

A direct measurement of U_v can be made if the isochrone line shown in Fig. 41(c) can be plotted from site measurements of excess pore pressure. This is demonstrated in Example 15.

Time factor T_v is a constant derived from the theory of consolidation and is found from the equation

$$T_v = \frac{c_v t}{d^2}$$

where: c_v = coefficient of consolidation;
   d = drainage path;
   t = time settlement has been taking place.

There is a relationship between T_v and U_v which depends on a number of variables such as distribution of load, drainage path, etc.

For practical purposes, however, the relationship between T_v and U_v may be taken as follows:

U_v	10	20	30	40	50	60	70	80	90
T_v	0·008	0·032	0·070	0·125	0·197	0·290	0·410	0·570	0·848

Up to $U_v = 60$ per cent this relationship gives a parabolic curve, the equation of which is $U_v{}^2 = \dfrac{4T_v}{\pi}$ or $U_v = 1.13\sqrt{T_v}$

Drainage path d. The drainage path may be loosely defined as the longest distance the water would have to travel to exit from the soil. It depends on the soils bounding the stratum being investigated. If the soil can drain at one face only, then d equals the full thickness of the stratum. If drainage is at both faces of the soil then d equals half the thickness of the stratum.

In the oedometer test, with porous stones top and bottom of the sample, the drainage path is always half the thickness of the sample.

EXAMPLE 15

Piezometers are installed at 0·5 m vertical intervals in a 3 m thick clay layer prior to construction of an embankment immediately above it. The embankment is 8 m high and a compacted clay soil with an average bulk density of 1800 kg/m² is used in its construction. One year after completion, the piezometers showed the following excess pore-water pressures:

Depth from top of clay (m)	0	0·5	1·0	1·5	2·0	2·5	3·0
Excess pore-water pressure (kN/m²)	130	124	115	101	82	53	0

It is observed that the embankment has settled 15 mm due to the consolidation of the underlying clay.

Estimate how long it will take for 90 per cent of the consolidation to be completed, and how much the final settlement will be.

If a free draining sand blanket had been laid at the base of the embankment, how much consolidation would have occurred one year after completion?

SOLUTION

Figure 42 is a graph of depth against *excess* pore water pressure. The curve plotted represents the isochrone for a time of one year after completion. The isochrone for a time immediately after completion (ignoring construction time), is represented by the vertical line at 144 kN/m² excess pore-water pressure. The isochrone at total final consolidation is represented by the vertical line at 0 kN/m² excess pore-water pressure.

The total consolidation that takes place is therefore represented by the whole area of Fig. 42.

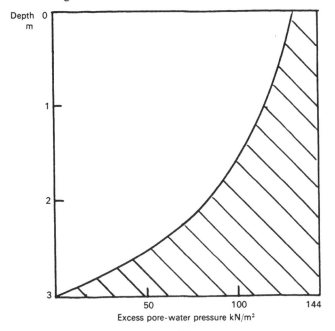

FIG. 42

After one year the isochrone measured by piezometers is the curve plotted, and the consolidation that has taken place after one year is represented by the shaded area in the figure.

$$\therefore \; U_v = \frac{\text{Consolidation after 1 year}}{\text{Total final consolidation}} \times 100 = \frac{\text{Shaded area of figure}}{\text{Total area of figure}} \times 100$$

graphical measurement shows this to be approximately 37 per cent.

For $U_v = 37\%$, $T_v = 0\cdot12$ (interpolated from $U_v - T_v$ relationship)

Inspection of Fig. 42 shows that the underlying stratum is relatively free-draining, whereas the embankment constructed is relatively impermeable. Therefore the drainage path is the full depth of the stratum, i.e. 3 m.

$$T_v = \frac{c_v t}{d^2}$$

$$0.12 = \frac{c_v \times 1}{3^2}$$

$$c_v = 1.08 \text{ m}^2/\text{year},$$

assuming c_v to be constant,

for 90% settlement $\qquad U_v = 90\% \qquad T_v = 0.848$

$$\therefore \qquad 0.848 = \frac{1.08 \times t_{90}}{3^2}$$

$$t_{90} = \underline{\underline{7 \text{ years}}}$$

Final settlement $= 15 \times \dfrac{100}{37} = \underline{\underline{40.5 \text{ mm}}}$

If a sand blanket had been laid the drainage path would be reduced to half the stratum thickness, and then after one year

$$T_v = \frac{1.08 \times 1}{1.5^2} = 0.48$$

for which $U_v = 76\%$

Settlement after one year $= \dfrac{76}{100} \times 40.5 = \underline{\underline{30 \text{ mm}}}$

Determination of the coefficient of compressibility

The coefficient of compressibility has been defined as the decrease in unit volume per unit increase of pressure

$$m_v = -\frac{1}{V_0} \frac{\delta V}{\delta p}$$

but since the soil is assumed not to expand in a lateral direction, i.e. the cross-sectional area remains constant, this may be expressed in terms of stratum thickness h,

i.e. $\qquad\qquad m_v = \dfrac{1}{h_0} \dfrac{\delta h}{\delta p} \qquad\qquad (1)$

also since the solids are assumed incompressible, any change in volume of the soil must be entirely due to a change in voids,

therefore $\qquad\qquad m_v = -\dfrac{1}{1 + e_0} \dfrac{\delta e}{\delta p} \qquad\qquad (2)$

$(1 + e_0)$ representing the total original volume *see* Chapter 1, page 2.

Therefore if a graph of thickness h after 24 hours against pressure p is plotted from the oedometer test results the slope of this graph will be $-\dfrac{\delta h}{\delta p}$ and m_v may be found from equation (1).

However, it is more usual to use equation (2) and plot a graph of void ratio against pressure.

From the test result e_f, the final void ratio, and h_f, the final thickness, are known.

$$Ah_f = 1 + e_f$$

but any change in volume must be a change in voids or $A\delta h = \delta e$ since the solids are incompressible and the cross-sectional area constant. Dividing one equation by the other,

$$\frac{\delta h}{h_f} = \frac{\delta e}{1 + e_f}$$

$$\delta e = \frac{1 + e_f}{h_f}\delta h$$

Since $(1 + e_f)/h_f$ is a constant and δh has been recorded, the void ratio e should now be calculated and plotted against effective pressure to give a compression curve.

From the definition of the coefficient or compressibility (Eqn 2)

$$m_v = -\frac{\delta e}{\delta p} \times \frac{1}{1 + e_0}$$

or m_v is the slope of the compression curve divided by the initial volume of the sample.

EXAMPLE 16

The results of a consolidation test on a clay sample are as follows:

Thickness of sample after 24 hours (mm)	19·92	18·75	17·94	17·38	17·06	16·92	18·46
Pressure (kN/m²)	0	100	200	300	400	500	0

The final moisture content was determined as 28 per cent and the grain specific gravity as 2·68.

Plot a graph of void ratio against pressure and determine the coefficient of compressibility between 250 to 380 kN/m² range of pressure.

SOLUTION

Final void ratio $e_f = 0.28 \times 2.68 = 0.7504$

$$\frac{1 + e_f}{h_f} = \frac{1 + 0.7504}{18.46} = 0.0948$$

p'	h	δh	δe	e
(kN/m^2)	(mm)	$(h_1 - h_2)$	$(0.095\delta h)$	
0	19.92			0.8888
100	18.75	−1.17	−0.1109	0.7779
200	17.94	−0.81	−0.0768	0.7011
300	17.38	−0.56	−0.0531	0.6480
400	17.06	−0.32	−0.0303	0.6177
500	16.92	−0.14	−0.0133	0.6044
0	18.46	+1.54	+0.1460	0.7504

Note: In the final column, the value $e = 0.7504$ is known. The other values of e are found by algebraically subtracting δe successively up the column. The graph of pressure against void ratio is shown in Fig. 43.

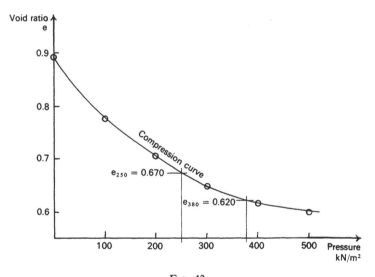

FIG. 43

Between the pressure range 250–380 kN/m²

$$e_{250} = 0.670 \qquad e_{380} = 0.620$$
$$m_v = \frac{0.670 - 0.620}{380 - 250} \times \frac{1}{1 + 0.670}$$
$$= 0.23 \times 10^{-3} \text{ m}^2/\text{kN}$$

Determination of coefficient of consolidation c_v

From the oedometer test results select the square root of time against compression graph applicable to the pressure range in use.

The determination of c_v from this graph may be carried out using various methods, the most common of which is demonstrated in Example 17.

EXAMPLE 17

A standard consolidation test carried out on a soil sample initially 20 mm thick gave the following results over a pressure range of 100–200 kN/m².

Time (minutes)	$\frac{1}{4}$	1	$2\frac{1}{4}$	4	9	16	25	36
Thickness of sample (mm)	19·82	19·64	19·50	19·42	19·28	19·12	18·98	18·84
Time (minutes)	49	64	81	100	121	144	169	196
Thickness of sample (mm)	18·68	18·54	18·40	18·28	18·20	18·10	18·04	17·99

After 24 hours the sample thickness was 17·61 mm

(a) Plot the graph of compression against square root of time and show which part of the curve represents consolidation settlement.

(b) Estimate the coefficient of consolidation for this soil.

(c) If the coefficient of compressibility was 0·00011 kN/m², estimate the coefficient of permeability.

(d) How long would it take for a 3-m-thick stratum of this soil, drained at both top and bottom surfaces, to reach 50 per cent of its total consolidation?

SOLUTION

(a) For the plotted graph see Fig. 44. There is a long section of the plotted curve (AB) which gives a straight line. The "kink" at the start of the curve is due to oedometer deflection as well as elastic settlement of the clay, and therefore a correction is applied by producing BA back to the vertical axis at C and considering this point as zero consolidation.

A line drawn through C such that the ratio PQ : PR = 1 : 1·15 (see Fig. 44) will cut the plotted curve at 90 per cent consolidation. Having located 0 and 90 per cent consolidation on the vertical axis, a linear scale is plotted and 100 per cent consolidation marked. The consolidation settlement from 0 to 100 per cent is marked on Fig. 44. The small section from 100 per cent consolidation to the actual 24-hour figure read in the test is said to be the secondary settlement.

(b) At 90 per cent consolidation (see Fig. 44)
$$\sqrt{t_{90}} = 12·3$$
since
$$T_v = \frac{c_v t}{d^2}$$
at 90 per cent U_v, $T_v = 0·848$
$$\therefore \quad 0·848 = \frac{c_v \times 12·3^2}{10^2}$$
$$c_v = 0·56 \text{ mm}^2/\text{min}$$
$$= 0·93 \times 10^{-9} \text{ m}^2/\text{s}$$

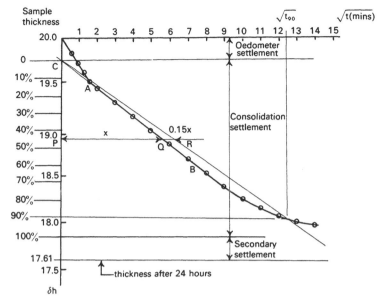

FIG. 44

(c) If
$$m_v = 0.00011 \text{ m}^2/\text{kN}$$
$$c_v = \frac{k}{m_v \gamma_w}$$
$$k = 0.93 \times 10^{-9} \times 0.00011 \times 9.8$$
$$= \underline{\underline{0.001 \times 10^{-6} \text{ mm/s}}}$$

(d) For 50 per cent consolidation of the stratum
$$T_v = 0.197$$
$$0.197 = \frac{0.93 \times 10^{-9} t_{50}}{1.5^2}$$
$$t_{50} = 0.4766 \times 10^9 \text{ s}$$
$$= \underline{\underline{15 \text{ years}}}$$

Settlement due to consolidation

Sands and other permeable strata may be considered to settle during the construction period. Consolidation of clays and silts may continue for years. Settlement due to consolidation is unlikely to be uniform, owing to uneven pressure distribution and variations in the soil both laterally and at depth.

Consolidation tests

Attempts to predict the amount and rate of settlement in soils of low permeability may be based on load tests, but these are not satisfactory. Results of consolidation tests are a better guide. For example:

Assume an increase of effective pressure δp.

Let m_v = Mean coefficient of compressibility for range of δp (found from consolidation test).

Decrease in unit volume = $m_v \delta p$.

Assuming no lateral strain:

Total reduction in thickness = $m_v \times \delta p \times h$

or more directly:

Decrease in unit volume = $\dfrac{e_0 - e_1}{1 + e_0}$

Reduction in thickness = $\dfrac{e_0 - e_1}{1 + e_0} h$

Settlement $\rho_c = m_v \delta p h = \dfrac{e_0 - e_1}{1 + e_0} h$

Rate of settlement

$$T_v = \frac{c_v t}{d^2}$$

For both stratum and sample T_v and c_v will be constant over a given range of pressure.

$$\therefore \quad \frac{t_1}{t_2} = \frac{d_1{}^2}{d_2{}^2}$$

where d is the length of drainage path which may be taken as $\frac{1}{2}h$ if drainage is both ways and h if drainage is one way only.

EXAMPLE 18

A rigid foundation block, circular in plan and 6 m diameter rests on a bed of compact sand 5 m deep. Below the sand is 1·6 m of clay overlying impervious bedrock. Groundwater level is 1·5 m below the surface of the sand. The density of the sand above G.W.L. is 1920 kg/m³, the saturated density of sand is 2080 kg/m³ and the saturated density of the clay is 1990 kg/m³.

A laboratory consolidation test on an undisturbed sample of the clay, 20 mm thick and drained top and bottom, gave the following results:

Pressure (kN/m²):	50	100	200	300	400
Void ratio:	0·73	0·68	0·625	0·58	0·54

If the contact pressure at the underside of the foundation is 200 kN/m²:

(a) Estimate the final average settlement of the foundation, assuming that load spread may be taken as 1 horizontal to 2 vertical.

(b) If the consolidation test sample reached 90 per cent consolidation in 1 hour 46 minutes, how long will it take the foundation to reach 90 per cent of its final settlement?

SOLUTION
Referring to Fig. 45(a):

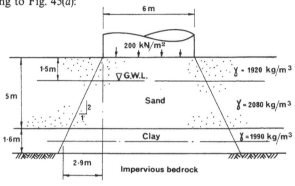

FIG. 45(a)

(a) Plot e against p (see Fig. 45(b)).
Initial effective pressure at centre line of clay

$$= (1\cdot5 \times 1920 + 3\cdot5 \times 2080 + 0\cdot8 \times 1990 - 4\cdot3 \times 1000)\frac{9\cdot8}{1000}$$
$$= 73 \text{ kN/m}^2$$

Final effective pressure at centre line of clay

$$= 73 + 200 \times \frac{6^2}{11\cdot8^2} = 124\cdot7 \text{ kN/m}^2$$

From Fig. 45(b) (the e against p graph):
$$e_o = 0\cdot699 \qquad e_f = 0\cdot665$$

Settlement

$$= \frac{0\cdot699 - 0\cdot665}{1 + 0\cdot699} \times 1600 = \underline{\underline{32 \text{ mm}}}$$

(b) Length of drainage path for sample $= 10 \text{ mm } (h/2)$
$$\text{site} = 1600 \text{ mm } (h)$$

$$\frac{106}{10^2} = \frac{t_{\text{site}}}{1600^2}$$

$$t_{\text{site}} = \frac{1600^2 \times 106}{60 \times 24 \times 365 \times 10^2}$$

$$= \underline{\underline{5\cdot2 \text{ years}}}$$

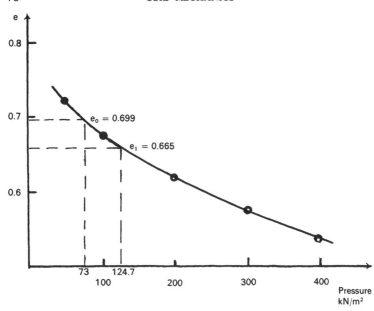

FIG. 45(*b*). $e \longrightarrow p$ graph

SETTLEMENT IN SANDS

Coarse-grained soils such as sand or gravel present a very different problem in the prediction of settlement. Firstly, owing to their poor cohesive properties, it is difficult to obtain an undisturbed sample of these soils for laboratory testing. More important however, the high permeability of coarse-grained soils allows consolidation to occur very rapidly. This means that it is difficult to differentiate between initial, elastic and consolidation settlement.

For these reasons the prediction of settlement in sands is always based on the results of tests carried out on site. The most common tests are the plate loading test, the cone test and the standard penetration test. Prediction of settlements from these three tests is empirical.

Plate loading test
In this test a steel plate is placed at foundation level and loaded to the same intensity of pressure as the final foundation. The settlement of the plate is then measured.

The major problem with the plate loading test is that the soil is only tested to a depth of about $1\frac{1}{2}$ times the breadth of the test-plate, whereas the foundation will stress the soil to a depth of $1\frac{1}{2}$ times the breadth of the foundation. If the soil is homogeneous to this greater depth then a reasonable correlation is obtained and the prediction of settlement may be based on the results. However, homogeneous soils are exceptional. The water table commonly gives rise to such a variation with depth, and if the water table falls within $1\frac{1}{2}$ times the breadth of the foundation, plate loading tests can be very misleading.

Plate loading tests are also relatively expensive to carry out and, in any event, may only be carried out at levels near the surface. Thus for borehole measurements either the cone test or the standard penetration test is used.

Cone test

This uses a standard tool consisting of a 60° cone of 10 cm² maximum cross sectional area. The cone is pushed into the soil with a hydraulic jack and the pressure required to overcome the resistance of the soil, the cone resistance C_{KD}, recorded. This test may be carried out in a borehole, and the value obtained for C_{KD} gives a direct value for end bearing capacity of piles. The estimation of settlement from cone resistance is also largely empirical and is based on past experience of this test.

One approach to the estimation of settlement in sands is based on consolidation theory and can be found from the expression.

$$\text{settlement} = \frac{h \times p_o'}{1 \cdot 5 \, C_{KD}} \log_e \frac{p_o' + \delta p}{p_o'}$$

where h = thickness of sand below the cone
            p_o' = effective overburden pressure at that depth
            δp = increase in pressure at that depth due the building load
    C_{KD} = cone resistance

If the sand stratum is relatively thick cone tests are carried out at regular intervals in the borehole. The interval used then becomes the thickness of sand below the cone (h) and the settlement of this layer is calculated. The total settlement is the summation of settlement for all these layers. The depth to which testing needs to be carried out generally depends on the value of the increase of pressure at that depth due to the building load (δp) and the variation of δp with depth is discussed in Chapter 7.

Standard penetration test

This test, described in detail in B.S. 1377 (1975), is the simplest to carry out and therefore, arguably, the most commonly used. However, it is also the most empirical of the three site tests.

The standard tool consists of a steel tube 800 mm long of 50 mm external diameter and 35 mm internal diameter. The lower end is formed into a cutting edge and the upper end can be attached to boring rods for use down a bore hole. In order to ensure it is penetrating undisturbed soil the tool is driven into the floor of the borehole to a depth of approximately 150 mm. A standard 64 kg drop hammer falling from a height of 750 mm is then used to drive the tube into the ground for a further 300 mm. The number of blows of the standard hammer required to drive the final 300 mm is known as the Standard Penetration Number, N.

The value of N enables an experienced engineer to make empirical judgements concerning the denseness, shear strength and likely settlement of the soil. A low value of N, say 5 to 10 blows, indicates a loose sand with low shear strength and high settlement whereas a high value, say 30 to 50 blows, would indicate a compact sand.

Many graphs and charts are available relating N to bearing capacity, angle of friction and settlement. A simple conversion of N to cone value C_{KD} is also used ($C_{KD} = KN$). C_{KD} is in kg/cm^2 and K varies from 2·5 to 16 depending on the grain size of the soil with a value of $K = 4$ commonly used for medium sand. All such conversions should however be treated with caution.

QUESTIONS

1. The following readings were obtained during one stage of a consolidation test on a sample of saturated clay, initially 20 mm thick, with drainage from both top and bottom faces of the sample:

Time (minutes)	Reduction in thickness (μm)	Time (minutes)	Reduction in thickness (μm)
0	0	9	432
0·25	109	16	525
1	172	25	610
2·25	236	36	665
4	302	49	687
6·25	361	24 hours	770

Plot the curve of reduction in thickness against square root of time, and from it determine the coefficient of consolidation of the clay in square metres per day.

2. Below the foundation of a structure there is a stratum of compressible clay 6 m thick with incompressible porous strata above and below. The effective overburden pressure at the centre line of the stratum before construction was 108 kN/m^2. After completion of the structure the pressure was increased by 170 kN/m^2. Oedometer tests were carried out on a sample

of the clay initially 20 mm thick. Each pressure was allowed to act for 24 hours and the decrease in thickness measured, the results being as follows:

Pressure (kN/m^2)	Thickness (mm)
0	10
50	19·70
100	19·41
200	18·99
400	18·52

Under a pressure of 200 kN/m^2, 90 per cent of the total consolidation took place in 21 minutes. Find:

(a) The probable settlement of the structure and
(b) The time in which 90 per cent of this settlement may be expected to occur.

3. The raft foundation of a very wide building is to be at a depth of 2·5 m below ground level. Borings show that a 4·5 m-thick stratum of clay, of average density 1800 kg/m^3, lies under sandy silt, 2·5 m deep, so that the foundation is in contact with the top of the clay layer. G.W.L. is 2 m below ground level; the average density of the sandy silt may be taken as 1920 kg/m^3 both above and below G.W.L. The estimated contact pressure at the underside of the foundation is 120 kN/m^2. A consolidation test is carried out, yielding the following results:

Applied pressure (kN/m^2):	25	50	100	150	200
Void ratio:		1·163	1·148	1·130	1·117	1·109

Calculate the effective pressure at the centre-line of the clay before and after construction, and hence estimate the final settlement of the raft.

4. The following readings were obtained in a consolidation test:

Pressure (kN/m^2) 0 50 100 200 400 600 800 0
Dial guage
after 24 hrs
(mm)   0 1·49 2·55 3·47 4·18 4·49 4·60 2·99

The initial thickness of the sample was 20 mm. If the final moisture content was 26 per cent and $G_S = 2·7$, plot a graph of void ratio against effective stress and find the value of the coefficient of compressibility for a pressure range from 140 to 380 kN/m^2.

If the coefficient of consolidation is $0·8 \times 10^{-3}$ $m^2/minute$, estimate the coefficient of permeability for this soil.

5. A clay layer, at depth below the water table, is subjected to a rapidly applied pressure increment of 200 kN/m^2 uniformly throughout the layer.

Piezometers placed at vertical increments in the clay showed the following excess pore-water pressures, one year after application of the load:

Depth in clay layer (m)	Top of clay layer	1 m	2 m	4 m	6 m	8 m	Bottom of clay 10 m
Excess pore-water pressure (kN/m^2)	0	70	110	148	144	110	0

From a consolidation test the coefficient of compressibility m_v was found to be 3.8×10^{-5} m^2/kN. Using the relationship between time factor T_v and degree of consolidation U_v given, estimate the time required for 600 mm consolidation to occur. Estimate the coefficient of permeability of the clay.

CHAPTER 6

SHEAR STRENGTH

The shear strength of a soil may be defined as the maximum resistance of the soil to shearing stress *under any given conditions.*

The conditions referred to above are mainly concerned with the drainage properties of the soil. For a coarse-grained soil, drainage is normally good and occurs as the test proceeds. A fine-grained soil, however, will drain very slowly, and therefore the rate of testing is an important factor.

There are several basic shear tests which may be carried out on a soil, which, if carried out with the same drainage conditions, should give comparable results.

SHEAR-BOX TEST

A simple test for finding the shear strength of a soil is the shear-box test (*see* Fig. 46). A shearing force is applied to the sample such that the sample shears at a constant rate of strain. The shearing resistance is measured on a proving ring, and the maximum value is

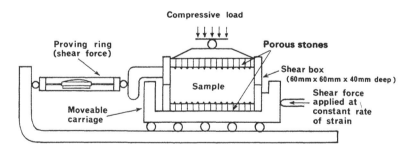

Fig. 46. Shear-box

the shear strength of the soil at failure. This shear strength may be found with the sample subjected to varying compressive loads and a graph of shear stress against compressive stress plotted, normally giving a straight line graph (*see* Fig. 47).

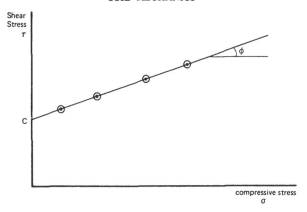

FIG. 47. Shear-box results

COULOMB'S LAW

The equation of the line obtained in this test is known as Coulomb's law and in its simplest form may be stated as:

$$\tau = c + \sigma \tan \varphi$$

where: τ = shear stress;

     σ = total compressive stress;

     c = apparent cohesion; and

     φ = angle of shearing resistance (internal friction).

It should be noted that although c and φ are called "apparent cohesion" and "angle of shearing resistance", they are only empirical constants for a particular soil under certain conditions of drainage, void ratio, etc. Laboratory values are of little use unless these conditions are the same on site.

EXAMPLE 19
A shear-box test carried out on a sandy clay gave the following results:

Vertical load (kg)	Divisions of proving ring dial gauge (one division to 1 μm)
36·8	17
73·5	26
110·2	35
146·9	44

If the shear box is 60 mm square and the proving ring constant 20 N/μm, determine the apparent cohesion and the angle of internal friction for this soil.

SOLUTION

Vertical load (W)	Compressive stress $\dfrac{W}{0.06^2 \times 1000}$ kN/m²	Dial gauge (S)	Shear stress $\dfrac{20S}{0.06^2 \times 1000}$ kN/m²
36·8	100	17	94
73·5	200	26	144
110·2	300	35	194
146·0	400	44	244

From the graph (Fig. 48),

$$\text{Apparent cohesion } c = \underline{\underline{47 \text{ kN/m}^2}}$$

$$\text{Angle of internal friction } \varphi = \underline{\underline{26°}}$$

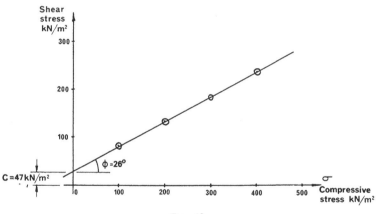

FIG. 48

CHECK:

$$\tau = c + \sigma \tan \varphi$$

$$194 = 47 + 300 \tan \varphi$$

$$\tan \varphi = \frac{147}{300} = 0.49$$

$$\varphi = 26°$$

Although the shear-box test gives a simple demonstration of Coulomb's law, it is not much used as a practical test. It is difficult to carry out on an undisturbed sample, and the distribution of stress is uncertain. The most common test on soils for determination of shear is the tri-axial test.

TRI-AXIAL COMPRESSION TEST

The tri-axial test is the most common method used in soil mechanics laboratories for finding the shear strength of a soil.

The soil specimen is extruded from a 37·5 mm-diameter cutting tube, capped top and bottom, and covered with a rubber membrane to prevent loss of moisture.

The prepared sample is placed in position (*see* Fig. 49) and the transparent cylinder filled with water. A measured pressure head is applied to the water, and the soil sample is then in similar conditions to the site conditions, where this lateral pressure would be due to the surrounding soil. This lateral pressure or cell pressure will be the minimum principal stress.

A vertical load is now applied to the sample at a constant rate of strain until the sample fails. This vertical applied pressure at failure, the deviator stress, may be measured on a proving ring, and when added to the cell pressure gives the maximum principal stress.

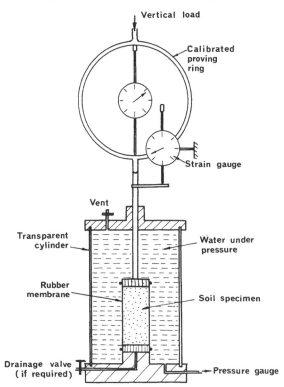

FIG. 49. Tri-axial compression apparatus

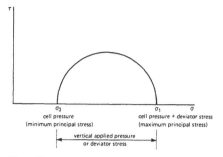

FIG. 50. Mohr circle for principal stresses

With the maximum and minimum principal stresses at the point of failure of the soil known, a Mohr circle can be drawn (*see* Fig. 50).

If a series of these tests are carried out with different cell pressures (σ_3) a series of Mohr circles may be drawn. These circles should have a common tangent known as the *Mohr envelope* (*see* Fig. 51), and will be the same as the line given by the Coulomb equation, provided the tri-axial and shear-box tests are carried out under similar conditions of drainage.

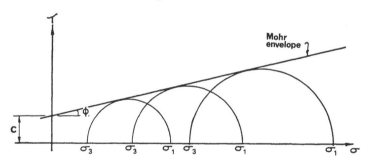

FIG. 51. Mohr envelope

EXAMPLE 20
The following data refer to three tri-axial tests performed on representative undisturbed samples of a soil:

Test	Cell pressure (kN/m^2)	Axial load dial reading (divisions) at failure
1	50	66
2	150	106
3	250	147

Load dial calibration factor 1·4 N per division. Each sample is 75 mm long and 37·5 mm diameter. Find by graphical means the value of apparent cohesion and the angle of internal friction for this soil.

SOLUTION

$$\text{Cross sectional area of sample} = \frac{\pi \times 37 \cdot 5^2}{4} = 1104 \text{ mm}^2$$

Cell pressure $(kN/m^2$	Additional vertical pressure (kN/m^2)	Total vertical pressure (kN/m^2)
50	84·4	134
150	134	284
250	186	436

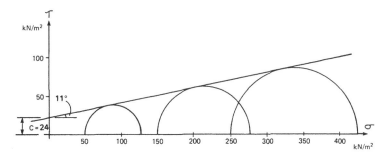

FIG. 52. Mohr envelope.

From Fig. 52:

$$\text{Apparent cohesion} = 24 \text{ kN/m}^2$$
$$\text{Angle of internal friction} = 11°$$

The tri-axial test may be carried out under different conditions of drainage, and the type of test will depend upon the site conditions and type of engineering works being undertaken.

An undrained test does not allow for drainage of the sample during the test. Since drainage is not required, not only need there be no provision for the water in the sample to drain but it is also unnecessary to allow *time* for drainage to occur. This test may therefore be carried out very speedily and is sometimes referred to as the "quick" test. If the sample is saturated an undrained test should give a value of φ equal to zero. This is, however, due to the condition of testing and not a property of the soil. The undrained test is a *total stress analysis* and is used where short-term values of shear strength are required such as for temporary excavations. It is also the standard test for bearing capacity of foundations which is a "short-term" case, since after initial loading the soil will consolidate and gain in shear strength (settlement is normally the main problem with foundations rather than bearing capacity).

A *drained test* must allow drainage of the sample during the test. Therefore, not only must the sample have porous stones top and bottom and the drainage valve open but the load must also be applied slowly to allow *time* for drainage to occur. This test is therefore sometimes referred to as the "slow" test. The drained test is an *effective stress analysis*, since all pore pressures are allowed to dissipate, and it is used where long-term values of shear strength are required, such as motorway cuttings.

A *consolidated-undrained test* allows drainage whilst the sample consolidates under cell pressure but is then sheared rapidly under conditions of no drainage. In this test the change in pore-water pressure in the soil is measured during the undrained part of the test, and hence an *effective stress analysis* may be conducted. The results of a consolidated-undrained test with pore-pressure measurement would be similar to those of a drained test, but the test is carried out more speedily. The measurement of pore-pressure change and dissipation are also of use in certain engineering contracts.

c and Φ parameters with respect to effective stress

When pressure is applied to a fine-grained soil the load is initially carried by the pore-water, but is gradually transferred to the soil particles as the water pressure dissipates. In the long term therefore;

$$\text{Effective stress } \sigma' = \sigma - u$$

and Coulomb's equation may be written more correctly:

$$\tau = c' + (\sigma - u) \tan \varphi'$$
or
$$\tau = c' + \sigma' \tan \varphi'$$

In this equation c' is sometimes termed the true cohesion and is referred to effective stress instead of total stress; φ' is the "true" angle of internal friction. To measure these values it is necessary to keep void ratio, density, etc., constant during the test, and as this is difficult in practice, it is more usual to find the arbitrary constants c and φ.

EXAMPLE 21

Explain the terms *total pressure, effective pressure* and *pore-water pressure* as applied to saturated soil and show how these pressures are related to one another. What is the influence of pore-water pressure on the shear strength of soil?

A series of undrained tri-axial tests on samples of saturated soil gave the following results:

Lateral pressure (kN/m²)	100	200	300
Pore water pressure (kN/m²)	20	70	136
Principal stress difference at failure (kN/m²)	290	400	534

Find the values of the parameters c and φ:

(a) with respect to total stress;
(b) with respect to effective stress.

SOLUTION

For explanation of terms *see* text.

(a) Total stress analysis

$$\sigma_3 = 100 \qquad 200 \qquad 300$$
$$\sigma_1 = 390 \qquad 600 \qquad 834$$

Referring to Fig. 53(a):

$$c = \underline{\underline{60 \text{ kN/m}^3}} \qquad \varphi = \underline{\underline{22°}}$$

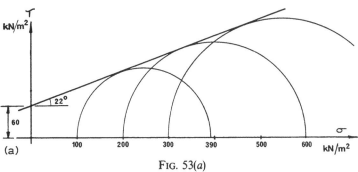

FIG. 53(a)

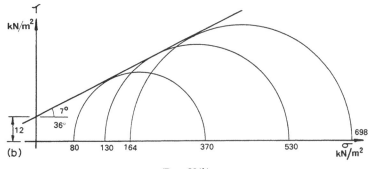

FIG. 53(b)

(b) Effective stress analysis

$$\sigma_3' = \quad 80 \qquad 130 \qquad 164$$
$$\sigma_1' = 370 \qquad 530 \qquad 698$$

Referring to Fig. 53(b):

$$c' = \underline{\underline{12 \text{ kN/m}^2}} \qquad \varphi = \underline{\underline{36°}}$$

FURTHER SHEAR TESTS

Unconfined compression test

For a rapid check of the shear strength of a cohesive soil on site the unconfined compression test may be used (*see* Fig. 54 and B.S. 1377). In this test the sample is placed between two discs without any lateral support. Vertical load is applied by hand through a spring, the strain of which is measured by an autographic recording arm on to a chart. Different stiffness springs may be used for different soil samples, and each spring has a transparent mask which when laid over the strain diagram gives the compressive stress on the sample.

In this test the sample is sheared rapidly and no drainage takes place. The lateral pressure is zero, and therefore the Mohr's circle would be as shown in Fig. 55.

If an unconfined test only is carried out there is no way of finding the angle of internal friction, since all tests should give the same circle. However, since this is an undrained test, normally on a saturated soil, the value of φ may be assumed to be zero. A value for the

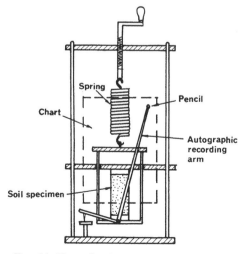

Fig. 54. Unconfined compression apparatus

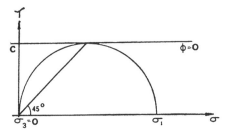

FIG. 55. Mohr envelope: unconfined compression test

shear strength of the soil can therefore be estimated, and it will be half the maximum compressive stress.

EXAMPLE 22

The following results were obtained from undrained shear-box tests on samples of silty clay:

Normal pressure (kN/m^2):   210   315   420
Shear strength (kN/m^2):   115   142   171

Find the apparent cohesion and the angle of shearing resistance. Find also the value of the apparent cohesion which would be expected from an unconfined compression test on a sample of the same soil.

If another specimen of this soil is subjected to an undrained tri-axial test with lateral pressure 280 kN/m^2, find the total axial pressure at which failure would be expected.

SOLUTION
From Fig. 56,

From shear box test: $c = 62$ kN/m^2; $\varphi = \underline{14°}$.

For unconfined compression test: $c = \underline{82}$ kN/m^2.

For tri-axial test: total axial pressure = $\underline{616}$ kN/m^2.

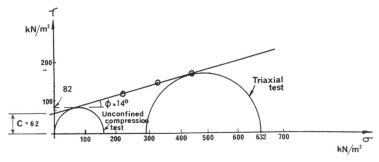

FIG. 56

All the examples in this chapter so far have been solved by graphical means. From the geometry of the Mohr circle (*see* Fig. 57) it can be shown that:

$$\sigma_1 = \sigma_3 N_\varphi + 2c\sqrt{N_\varphi}$$

where: σ_1 = maximum principal stress;

σ_3 = minimum principal stress;

c = apparent cohesion;

$N_\varphi = \tan^2 (45 + \varphi/2)$, where φ = angle of internal friction.

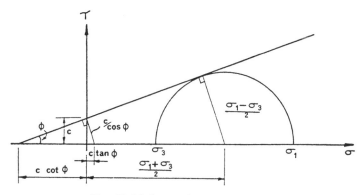

FIG. 57. Mohr envelope: general case

Suppose this equation were used in Example 22 for the tri-axial test, where:

$$c = 62 \text{ kN/m}^2 \qquad \varphi = 14° \qquad \sigma_3 = 280 \text{ kN/m}^2$$
$$\sigma_1 = 280 \tan^2 (52°) + 2 \times 62 \times \tan 52°$$
$$= 280 \times (1.28)^2 + 124 \times 1.28$$
$$= \underline{\underline{617 \text{ kN/m}^2}}$$

Vane test

This is a useful method of measuring the *in situ* shear strength of a clay. A cruciform vane is made up and the dimensions recorded (*see* Fig. 58 and B.S. 1377). This vane is driven into the soil at the bottom of a borehole and a measured torque applied at the surface. A cylinder of soil will resist this torque until at the moment of failure

$$\text{Torque } T = Px = c(\pi dh)\frac{d}{2} + c\left(\pi\frac{d^2}{4}\right)\tfrac{1}{3}d \times 2$$

(sides of   (ends of
cylinder)   cylinder)

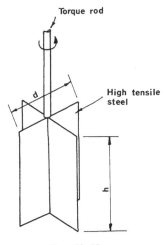

Torque rod

High tensile
steel

d

h

FIG. 58. Vane

since all the terms are known, c may readily be found. (Note φ assumed = zero).

This test is found to agree with tri-axial tests at depths between 15 and 30 metres.

The assumption that φ is zero is reasonable for a saturated soil and this test is found to agree with tri-axial tests at depths between 15 to 30 metres.

One difficulty with vane tests is that the soil does not fail on the perimeter of the vane but at some small distance beyond. Corrections for this overbreak can be made, but for most practical purposes may be assumed to be negligible. A more serious problem arises when the shear strength of the clay in the vertical plane is different from the shear strength in the horizontal plane. In this situation the test is carried out with two vanes with different length/diameter ratios, giving two equations and two values of c, i.e.,

$$T = c_v(\pi dh)\frac{d}{2} + c_H\left(\pi\frac{d^2}{4}\right)\tfrac{1}{3}d \times 2$$

The values of cohesion in the vertical plane (c_v) and in the horizontal plane (c_H) may then be calculated.

EXAMPLE 23

In a vane test a torque of 46 Nm is required to cause failure of the vane in a clay soil. The vane is 150 mm long and has a diameter of 60 mm. Calculate the apparent shear strength of the soil from this test.

When a vane of 200 mm length and diameter 90 mm is used in the same soil, the torque at failure was 138 Nm. Calculate the ratio of shear strength of the clay in a vertical direction to that in a horizontal direction.

SOLUTION

For original test only

$$46 \times 10^3 = \frac{c \times \pi \times 60^2 \times 150}{2} + \frac{c \times \pi \times 60^3}{6}$$

$$c = \underline{\underline{47 \cdot 9 \text{ kN/m}^2}}$$

For both tests

$$46 \times 10^3 = \frac{c_v \times \pi \times 60^2 \times 150}{2} + \frac{c_H \times \pi \times 60^3}{6}$$

$$138 \times 10^3 = \frac{c_v \times \pi \times 90^2 \times 200}{2} + \frac{c_H \times \pi \times 90^3}{6}$$

From these equations

$$c_v = 50 \cdot 2 \text{ kN/m}^2 \qquad c_H = 30 \text{ kN/m}^2$$

$$\therefore \quad \frac{c_v}{c_H} = \frac{50 \cdot 2}{30} = \underline{\underline{1 \cdot 67}}$$

Residual shear test

For long-term prediction of the stability of cuttings in clay a residual value of shear strength may be required. This test can be carried out using the shear box apparatus at a very low rate of strain. A saturated sample is tested continuously over a period of days, the shear box being reversed each time it has completed its full travel.

A graph is plotted showing shear resistance against strain. The typical form of the curve is shown in Fig. 59.

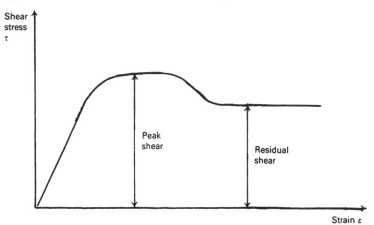

FIG. 59. Peak and residual shear

It can be seen that the shear resistance reaches a peak value and then drops to a residual value.

If a series of these tests are carried out at different vertical compressive loads, then the normal shear box graph of shear strength against compressive load may be plotted for both peak and residual values. An example of this is shown in Fig. 60.

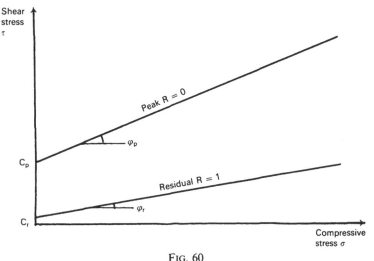

FIG. 60

The shear strength of a soil in a newly constructed cutting will be at peak value (s_p). With time this shear strength will dissipate and ultimately residual shear (s_r) will be reached. During the design life of a cutting therefore the actual shear strength of the soil ($\hat{s}$) will lie somewhere between these two values.

The residual value R is taken as:

$$R = \frac{s_p - \bar{s}}{s_p - s_r}$$

It can be seen that $R = 0$ immediately after construction and $R = 1$ after infinite time.

The residual factor may be used to predict a suitable design life for the cutting.

EXAMPLE 24

The results of a series of residual shear tests are as follows:

Compressive stress (kN/m²)	150	300	450
Peak shear stress (kN/m²)	144	212	280
Residual shear stress (kN/m²)	38	77	115

Estimate the shear parameters of the soil for a residual factor of 0·4.

SOLUTION
A graph of the results is shown in Fig. 61.
For a residual factor of 0·4 construct to scale an intermediate line between $R = 0$ and $R = 1$.

The shear parameters for this line may now be read from the graph, i.e.
$c = \underline{\underline{46 \text{ kN/m}^2}}, \varphi = 20°$

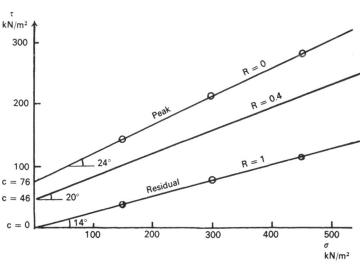

FIG. 61

PORE PRESSURE PARAMETERS

The stability of embankment slopes is more critical during construction than after a passage of time. As each layer of soil is compacted onto the embankment there will be an increase in pore-water pressure in the lower layers, with a consequent reduction in effective stress ($\sigma' = \sigma - u$). If therefore construction is too rapid, failure may occur. However, it would not be economic to allow time for these excess pore-water pressures to completely dissipate before adding another layer of soil. Therefore an attempt must be made to keep a check on the magnitude of these pore-water pressures.

The derivation of the simplified elastic expression for measuring these increases in pore-water pressure is beyond the scope of this book, but may be taken as:

$$\Delta u = B[\Delta\sigma_3 + A(\Delta\sigma_1 - \Delta\sigma_3)]$$

where: Δu = change in pore-water pressure

                 $\Delta \sigma_3$ = change in *total* lateral pressure

                 $\Delta \sigma_1$ = change in *total* vertical pressure

A and B are parameters based on practical observations.

To estimate the change in pore-water pressure therefore it is necessary to know the values of the parameters A and B.

The measurement of these parameters is carried out in a tri-axial apparatus arranged to facilitate measurement of pore-water pressure. A sample of the compacted soil is set up, a small cell pressure is applied and the sample allowed to consolidate. When consolidation is complete an increase in cell pressure $\Delta \sigma_3$ is applied and the increase in pore pressure Δu measured. The deviator stress $(\sigma_1 - \sigma_3)$ is not applied and therefore $(\Delta \sigma_1 - \Delta \sigma_3)$ is zero.

From the simplified expression therefore

$$\Delta u = B[\Delta \sigma_3 + A \times 0]$$

$$\text{or} \quad B = \frac{\Delta u}{\Delta \sigma_3}$$

which can be readily calculated.

If the soil is saturated then the increase in cell pressure is carried entirely by the pore water and B will be equal to 1.

Having determined the value of parameter B the excess pore-water pressure is allowed to dissipate. Measurement of the rate of dissipation at this stage gives a useful variation of the consolidation test and the coefficient of consolidation, c_v, can be determined.

When the excess pore-water pressure had completely dissipated a vertical stress σ_1 is applied and there will be another increase in pore-water pressure. This increase is measured throughout the test until failure of the soil sample. In this case there is no change in lateral pressure and $\Delta \sigma_3$ is zero. The expression for increase of pore-water pressure then becomes:

$$\Delta u = B[0 + A(\Delta \sigma_1 - 0)]$$

$$\text{or} \quad AB = \frac{\Delta u}{\Delta \sigma_1}$$

hence the value of AB may be determined and since B is known, parameter A may be found. The value of parameter A will vary with increasing load up to failure of the sample.

EXAMPLE 25

A sample of compacted soil 200 mm long and 100 mm diameter with drainage at each end is allowed to fully consolidate in a tri-axial apparatus under a cell pressure of 100 kN/m². The cell pressure is then increased to 250 kN/m² and an increase in pore-water pressure of 130 kN/m² is recorded. This pore-water pressure is allowed to dissipate and the rate of dissipation recorded as shown:

Pore-water pressure (kN/m²)	110	88	76	68	62	53	45	33	18	11
Time (hours)	0	1	2	3	4	6	8	12	18	24

The sample is then subjected to a deviator stress and pore-water pressure measurements taken at intervals until the sample fails in shear. The recordings of deviater stress, strain and pore-water pressure were as follows.

Strain (%)	0	2	4	6	8	10	
Deviator stress (kN/m²)	0	36	72	108	144	180	sample fails at 10% strain
Pore-water pressure (kN/m²)	0	10·8	35·3	67·6	100·8	128·4	

Determine the value of parameter B and the coefficient of consolidation. Plot a graph of parameter A against strain up to failure of the soil.

SOLUTION

$$\text{Parameter } B = \frac{\Delta u}{\Delta \sigma_3} = \frac{130}{150} = \underline{\underline{0 \cdot 87}}$$

For graph of pore-water pressure against time see Fig. 62(a)

From the graph, 50 per cent consolidation has taken place after 5·5 hours.
For 50% consolidation, $T_v = 0·197$

$$0 \cdot 197 = \frac{c_v \times 55 \times 60}{100^2}$$

$$c_v = 6 \text{ mm}^2/\text{min}$$

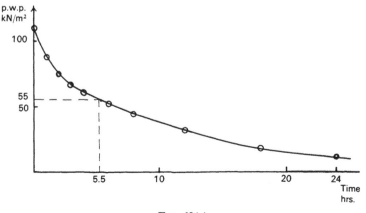

FIG. 62(a)

From results of the test to failure

Strain	0	2	4	6	8	10
Parameter $AB \left(\dfrac{\text{pore-water pressure}}{\text{deviator stress}} \right)$	0	0·30	0·49	0·63	0·70	0·71
Parameter A	0	0·34	0·56	0·72	0·80	0·82

These results are plotted in Fig. 62(b)

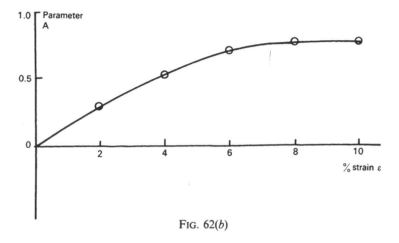

FIG. 62(b)

QUESTIONS

1. Undrained shear-box tests on samples of sandy clay gave the following results:

Normal pressure (kN/m²)	50	125	250
Shearing resistance (kN/m²)	61	83	119·5

Find the apparent cohesion and the angle of shearing resistance.

A tri-axial test is carried out on a sample of the same soil, with a cell pressure of 100 kN/m². Find the total axial stress at which failure would be expected.

2. Explain the following conditions for testing the shear strength of soils:

(a) Undrained.
(b) Consolidated undrained with pore pressure measurement.
(c) Drained.

Give examples of practical engineering problems in which these forms of testing would be used.

An undrained tri-axial test on a sample of compacted fill gave the following results:

Cell pressure (kN/m²)	100	200	300
Deviator stress (kN/m²)	191	243	296

What value of shear strength would have been expected if an unconfined compression test had been carried out on this soil?

3. Define effective stress and comment on its importance in practical soil mechanics problems.

From tri-axial tests with pore-water pressure measurement it is found that the cohesion and angle of shearing resistance of a soil, referred to effective stress, are 10 kN/m^2 and $25°$ respectively. Using Coulomb's equation find the shearing strength of this soil at a depth of 9 m below ground surface. The soil has an average density of 1930 kg/m^3 and the water table is at a depth of 2·7 m below the surface.

4. In an undrained tri-axial test on three specimens of sandy clay taken from a depth of 3 m below ground level the following results were obtained:

Cell pressure (kN/m^2)	Deviator stress (kN/m^2)
200	221
400	365
600	505

(a) Draw the Mohr diagram and determine the apparent cohesion and angle of shearing resistance of the soil.

(b) Derive Coulomb's equation for this soil.

(c) If a superimposed load of 60 kN/m^2 is applied to the top of the sandy clay layer, at ground level, find the permissible shear stress at a depth of 3 m, assuming the Coulomb equation remains unchanged and $\gamma = 2000 \text{ kg/m}^3$. The water table is below the 3 m level.

5. Enumerate the types of laboratory tri-axial test you would specify to be carried out in connection with the following field problems:

(a) The stability of the clay foundation of an embankment; the rate of construction being such that some consolidation of the clay occurs.

(b) The initial stability of a footing on saturated clay.

(c) The long-term stability of a slope in a stiff-fissured clay.

Give the reasons for your choice of test.

The table contains data obtained in consolidated–undrained tests on a clay soil. Determine the shear-strength parameters in terms of effective stress. A different specimen of the same soil is tested in consolidated–undrained tri-axial compression at a cell pressure of 400 kN/m^2, and fails when the deviator stress is 450 kN/m^2. Calculate the pore pressure in the specimen at failure.

Cell pressure during consolidation and shear (kN/m^2)	At failure	
	Deviator stress (kN/m^2)	Pore water pressure (kN/m^2)
300	260	120
800	615	330

6. A vane of length 250 mm and diameter 100 mm is used to measure the shear strength of a saturated soil. If the torque required to fail the vane is 518 Nm calculate the apparent shear strength of the soil.

A test on the same soil was carried out using a vane of 300 mm length and diameter 100 mm and the torque at failure was 612 Nm. Calculate the ratio of shear strength in the vertical plane to shear strength in the horizontal plane.

7. The following results were obtained in three residual shear tests:

Time (hours)	0	10	20	30	40	50	60	70	80	90	100	110	120	130	140
Shear stress (kN/m^2)															
Test 1	0	60	97	100	87	43	21	20	20						
Test 2	0	74	120	136	136	136	114	78	47	40	40	40			
Test 3	0	103	142	161	171	172	172	172	168	150	114	82	64	60	60

The compressive stresses were 100, 200 and 300 kN/m^2 in tests 1, 2 and 3 respectively. Estimate the residual factor for a time when the shear parameters will have values $c = 33$ kN/m^2, $\varphi = 18°$.

8. An embankment which has remained stable for 30 years is to be increased in height by adding 3 metres of soil, the existing side slopes to be maintained.

The compacted density of the additional soil is 1800 kg/m^3. Investigations show that any increase in vertical pressure causes an increase in lateral pressure such that

$$\frac{\text{Increase in lateral pressure}}{\text{Increase in vertical pressure}} = 0.4$$

Pore pressure parameters for the existing embankment are $A = 0.6$, $B = 0.9$. Calculate the increase in pore-water pressure in the existing embankment.

CHAPTER 7

DISTRIBUTION OF VERTICAL PRESSURE

When a pressure is applied to a soil, the distribution of stress within the soil mass is extremely complex. Simple elastic analysis of stress distribution assumes a semi-infinite, elastic, isotropic and homogeneous material, conditions which are most unlikely on any site. However, it is necessary to have some means of estimating the distribution of vertical pressure beneath a foundation, for instance, in the calculation of settlement. Boussinesq's elastic analysis gives a means of calculating this distribution and, for reasonably homogeneous clays, gives acceptable values for vertical pressure at depth.

SINGLE POINT LOAD (BOUSSINESQ'S THEORY)

Although a point load is unlikely for a foundation, the theory of pressure distribution beneath such a load provides a basis for the development of techniques concerning distributed loads.

Figure 63 shows a point load P acting on a soil mass. σ_v is the vertical stress at a point depth z below the point load and at horizontal distance r from its line of action.

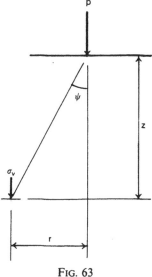

FIG. 63

Boussinesq's elastic analysis gives the relationship

$$\sigma_v = \frac{3P}{2\pi z^2} \left[\frac{1}{1 + (\frac{r}{z})^2} \right]^{5/2}$$

if r/z is taken as tan ψ (*see* Fig. 63) this equation can be written

$$\sigma_v = \frac{3P}{2\pi z^2} \cos^5\psi$$

From either of these equations it is a simple matter to calculate the distribution of stress at any points within the soil mass.

EXAMPLE 26
A point load of 700 kN is applied to a thick layer of clay. It is estimated that a pressure below 25 kN/m^2 will cause negligible settlement in the clay. Draw a bulb of pressure diagram, based on a 500 mm square grid, showing the limits of this pressure. Also show the distribution of pressure, due to the point load, along a horizontal section at depth 2 m.

SOLUTION
Since the question requires a 500 mm square grid which must extend to cover values down to 25 kN/m^2, the first step should be to determine the limits for calculation.

Inspection of Boussinesq's equation shows that for any given depth z the maximum value of σ_v will be when $r/z = 0$, i.e. $r = 0$ (directly beneath the point load as would be expected). Therefore, for a vertical pressure $\sigma_v = 25$ kN/m^2,

$$25 = \frac{3 \times 700}{2\pi z^2} \left[\frac{1}{1 + (\frac{0}{z})^2} \right]^{5/2}$$

$$z = 3.66 \text{ m}$$

i.e. the 500 mm grid need only be extended to a depth of, say, 4·0 m.

Value of r	0	0·5	1·0	1·5	2·0
For $z = 0.5$					
r/z	0	1·0	2·0		
σ_v	1337	236	23·9		
For $z = 1.0$					
r/z	0	0·5	1·0	1·5	
σ_v	334	191	59·1	17·5	
For $z = 1.5$					
r/z	0	0·33	0·67	1·0	
σ_v	149	114	59·2	26·2	11·6
For $z = 2.0$					
r/z	0	0·25	0·5	0·75	1·0
σ_v	83·6	71·8	47·8	27·3	14·8

For $z = 2.5$

r/z	0	0·2	0·4	0·6
σ_v	53·4	48·5	36·9	24·8

For $z = 3.0$

r/z	0	0·17	0·33	0·5
σ_v	37·1	34·7	28·5	21·3

For $z = 3.5$

r/z	0	0·14	0·29
σ_v	27·3	25·9	22·4

For $z = 4.0$

r/z	0	0·13
σ_v	20·1	20·1

These values are plotted on a grid as shown in Fig. 64. The bulb of pressure for 25 kN/m² vertical pressure can then be drawn by interpolating between the grid points. The distribution of vertical pressure along a horizontal plane at depth 2 m is also shown in the figure.

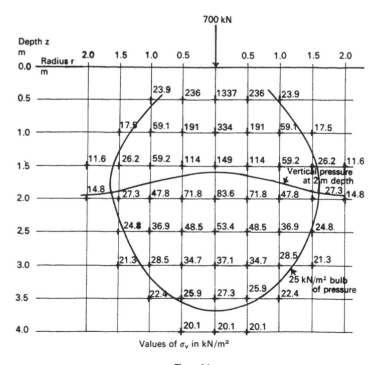

FIG. 64

DISTRIBUTED LOAD

In foundation design the load will normally be distributed. Although the initial pressure between a base and the soil can be complex, depending on factors such as the relative rigidity of the base and the soil reaction, it is usual to assume that it soon approximates to uniform distribution. Boussinesq's equation for point loading is then integrated over the area of the base to find the vertical pressure distribution at depth. This is a time-consuming process and the following two methods have been developed to simplify the task.

FADUM'S METHOD

Vertical stress under the corner of a rectangle

To find the vertical stress at point P, depth z beneath the corner of a flexible, rectangular base, Boussinesq's equation may be used. The load per unit area (q) is integrated over the whole rectangle for the point P in question. This integration is beyond the scope of this volume and produces a complex equation for σ_v in terms of length/depth and width/depth ratios. However, the value of σ_v may be found using Fadum's influence chart. If the length of the foundation is L, the breadth B and the depth to point P is z, then put $L/z = m$ and $B/z = n$. The chart shown in Fig. 65 may then be used to find an influence factor k. The vertical stress at point P, depth z beneath a corner of the rectangle, σ_v, is then equal to $k \times q$.

EXAMPLE 27

Calculate the distribution of vertical stress at 3 m vertical intervals, to a depth of 12 m, below the corner of a rectangular base length 6 m, width 4 m. The underside of the base exerts a uniform pressure of 120 kN/m² on the soil.

SOLUTION

Depth (metres)	$m = L/z$	$n = B/z$	k	σ_v (kN/m²)
3	2·0	1·33	0·218	26·2
6	1·0	0·67	0·147	17·6
9	0·67	0·44	0·088	10·6
12	0·5	0·33	0·062	7·4

Note that m and n are interchangeable.

This method may be readily developed to find the distribution of vertical stress beneath any point of a rectangular base.

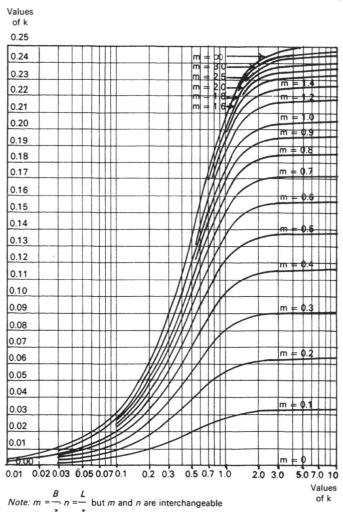

Values
of k

Note: $m = \dfrac{B}{z}$, $n = \dfrac{L}{z}$ but m and n are interchangeable

FIG. 65. Fadum's influence chart

EXAMPLE 28

An 8 m by 6 m flexible raft foundation for a building is to be constructed centrally on a 12 m × 9 m site such that a surround remains between the edge of the foundation and the boundary fence. The pressure at the underside of the foundation is 200 kN/m².

Plot the distribution of vertical pressure along a diagonal line across the site at depths of 2, 5 and 10 metres. Sketch the approximate bulb of pressure for 30 kN/m².

SOLUTION

Figure 66(a) shows the site (*ABCD*) and the foundation (*EFGH*). The diagonal *DB* has been chosen and the pressure distribution will be found under points *D* and *H* at the corners of the site and foundation respectively (points *B* and *F* will be identical). Line *HF* has been divided into four equal lengths to give points *K* (and *M*) and *L*

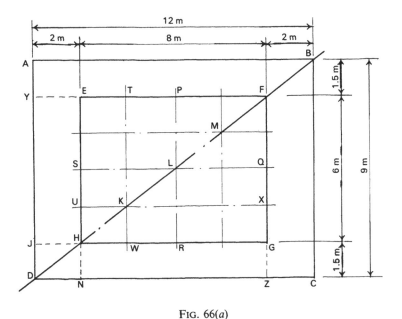

FIG. 66(a)

Point H (and F)

These are straightforward calculations of the pressure beneath a corner of the 8 m × 6 m foundation.

Depth (metres)	$m = L/z$	$n = B/z$	k	σ_v (kN/m²)
2	4	3	0·245	49
5	1·6	1·2	0·214	42·8
10	0·8	0·6	0·123	24·6

Point L

Since the raft is flexible it may be considered as four rafts *EPLS, PFQL, LQGR* and *SLRH* (*see* Fig. 66(a)). These areas are all of equal size and *each* has a corner at point L. Therefore if the distribution of pressure beneath a corner for one of these is found, the distribution beneath the centre of the complete raft will be the sum of all the areas with a corner at L, i.e. four times σ_v for one area.

Depth (metres)	$m = L/z$	$n = B/z$	k	σ_v (kN/m^2)	For whole raft $\Sigma\sigma_v = \sigma_v \times 4$ (kN/m^2)
2	2	1·5	0·223	44·6	178·4
5	0·8	0·6	0·123	24·6	98·4
10	0·4	0·3	0·048	9·6	38·4

Points K (and M)
For these two points the raft must be divided into four rectangles *ETKU*, *UKWH*, *TFXK*, *KXGW* and *UKWH* all with a corner at point K. The sum of the pressures below K for each of these areas will give the distribution beneath point K for the whole raft.

Depth (metres)	Area	Dimension (metres)	$m = L/z$	$n = B/z$	k	σ_v (kN/m^2)	For whole raft $\Sigma\sigma_v$ (kN/m^2)
2	*ETKU*	4·5 × 2	2·25	1	0·202	40·4	
	UKWH	2 × 1·5	1	0·75	0·155	31·0	155·2
	TFXK	6 × 4·5	3	2·25	0·242	48·4	
	KXGW	6 × 1·5	3	0·75	0·177	35·4	
5	*ETKU*	4·5 × 2	0·9	0·4	0·098	19·6	
	UKWH	2 × 1·5	0·4	0·3	0·048	9·6	81·8
	TFXK	6 × 4·5	1·2	0·9	0·180	36	
	KXGW	6 × 1·5	1·2	0·3	0·083	16·6	
10	*ETKU*	4·5 × 2	0·45	0·2	0·036	7·2	
	UKWH	2 × 1·5	0·2	0·15	0·013	2·6	33·2
	TFXK	6 × 4·5	0·6	0·45	0·085	17	
	KXGW	6 × 1·5	0·6	0·15	0·032	6·4	

Point D (and B)
In this case the point considered lies outside the loaded area. When this occurs, extend the loaded area *to have a corner at the point being considered* (D), i.e. consider a rectangle *YFZD* and find the pressure distribution beneath D. To deal with the unloaded areas simply take rectangles *with a corner at D*

and consider them with negative load, i.e. rectangles $YEND$ and $JGZD$ with – 200 kN/m² load.

In this case area $JHND$ has been considered twice as negatively loaded and must therefore be added as positive load.

The algebraic sum of all these pressure distributions gives the distribution beneath point D.

Depth (metres)	Area	Dimensions (metres)	Load (kN/m²)	$m = L/z$	$n = B/z$	k	σ_v (kN/m²)	For whole site $\Sigma\sigma_v$ (kN/m²)
2	YFZD	10 × 7·5	200	5	3·75	0·247	49·4	
	YEND	7·5 × 2	−200	3·75	1	0·204	−40·8	4·0
	JGZD	10 × 1·5	−200	5	0·75	0·178	−35·6	
	JHND	2 × 1·5	200	1	0·75	0·155	31·0	
5	YFZD	10 × 7·5	200	2	1·5	0·223	44·6	
	YEND	7·5 × 2	−200	1·5	0·4	0·109	−21·8	14·6
	JGZD	10 × 1·5	−200	2	0·3	0·089	−17·8	
	JHND	2 × 1·5	200	0·4	0·3	0·048	9·6	
10	YFZD	10 × 7·5	200	1	0·75	0·155	31·0	
	YEND	7·5 × 2	−200	0·75	0·2	0·050	−10·0	15·8
	JGZD	10 × 1·5	−200	1	0·15	0·040	− 8·0	
	JHND	2 × 1·5	200	0·2	0·15	0·014	2·8	

These distributions of vertical pressure are plotted in Fig. 66(b).

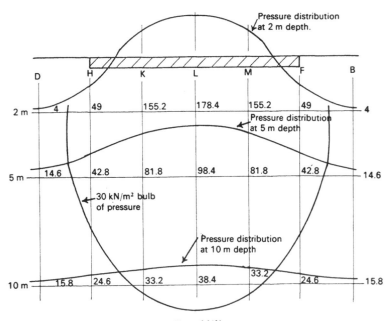

FIG. 66(b)

NEWMARK'S CHART

Vertical stress under the centre of a circular area

As with the stress under the corner of a rectangle, the stress under the centre of a circular loaded area may be found by integrating Boussinesq's basic equation.

For a circular area radius R this integration gives the result:

$$\sigma_v = q\left[1 - \left(\frac{1}{1 + (\frac{R}{z})^2}\right)^{3/2}\right]$$

Although the construction of circular foundations is uncommon, this equation enables us to construct an influence diagram.

Construction of Newmark's influence chart

The equation above may be re-arranged to give

$$\frac{R}{z} = \sqrt{\left[\left(1 - \frac{\sigma_v}{q}\right)^{-2/3} - 1\right]}$$

σ_v/q will vary from 0 to 1. Select a range of values at equal multiples between 0 and 1, and calculate the corresponding values of R/z, i.e.

σ_v/q	0·0	0·2	0·4	0·6	0·8	1·0
R/z	0·0	0·40	0·64	0·92	1·39	∞

Choose any convenient dimension for z (say 5 m) and calculate R

R	0	2·00	3·20	4·60	6·95	∞

Draw concentric circles of these radii to scale, ignoring the circle of infinite radius, as shown in Fig. 67.

There are now five areas; four annular rings (including the outside ring of infinite radius) and one central circle. *Each of these areas, if loaded with uniform load, would cause the same increment of vertical pressure at depth 5 m below the central point on the chart.*

Next, divide these areas into equal segments by drawing a number of rays at equal intervals. In Fig. 67 an angular increment of 30° has been chosen to give 12 segments. There are now 60 small areas each of which if loaded with uniform load, would cause the same increment of vertical pressure at depth 5 m below the central point. These small areas may be referred to as influence areas.

Any loading q applied to any part of the chart therefore, will cause an increase of vertical pressure σ_v at a point 5 m below the centre of the chart, where:

$$\sigma_v = q \times \frac{\text{number of influence areas covered by load.}}{\text{Total number of influence areas.}}$$

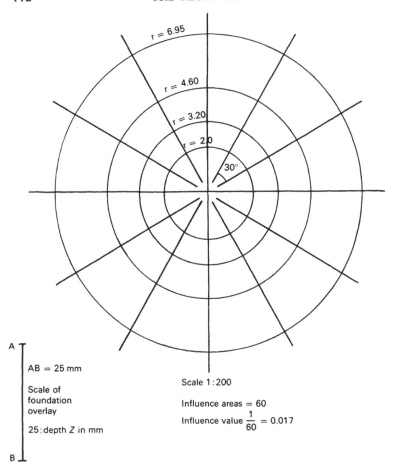

A

AB = 25 mm

Scale of
foundation
overlay

25 : depth Z in mm

B

Scale 1 : 200

Influence areas = 60

Influence value $\dfrac{1}{60}$ = 0.017

FIG. 67. Newmark's influence chart

If the pressure at depth 5 m below any point P, caused by a
foundation load, is required, this influence diagram may be used.
Draw the foundation on tracing paper to the same scale as the
chart, and lay over the chart such that point P on the foundation is
at the centre. The number of influence areas covered by the
foundation may now be counted (making suitable estimates for part
areas) and σ_v calculated.

The chart may be used for depths other than 5 m by varying the
scale of the foundation drawing. To facilitate this, draw to the same
scale as the circles, a line on the chart, of length equal to the depth z
selected. In Fig. 67, with scale 1 : 200 and depth z equal to 5 m, the

line AB will be 25 mm long. Use this line AB to calculate the scale required for the foundation drawing, i.e.:

For depth z	Scale of foundation drawing
4	25:4000 or 1:160
5	25:5000 or 1:200
8	25:8000 or 1:320
10	25:10000 or 1:400
20	25:20000 or 1:800
etc.	

The value of σ_v is calculated for the new depth exactly as before.

Note: At this point it is strongly recommended that readers should draw their own influence chart. The limitations of page size in this book has necessarily reduced accuracy and produced unusual scales for the foundation drawings. Example 29 and the questions at the end of this chapter are based on a chart with values of σ_v/q equal to 0·1, 0·2, 0·3 etc., a chosen z value of 4 m and rays drawn at 18° intervals. Such a chart drawn to a scale of 1:100 fits neatly onto A4 size paper, and gives 200 influence areas. Once drawn, such a chart can be used for any pressure distribution calculations.

EXAMPLE 29

Fig. 68(a) shows the layout of two raft bases for adjacent structures, and the pressure at the underside of each base. Calculate the distribution of vertical pressure under point P due to this foundation, at depths of 4, 8, 12 and 20 m.

Check the results using the method of distribution beneath the corner of a rectangle.

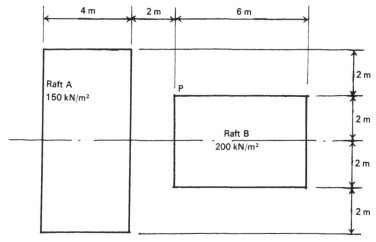

FIG. 68(a)

SOLUTION

This solution has been based on a Newmark chart drawn as recommended in the previous paragraph. The inner circles of the chart are drawn to scale in Fig. 68(b) with the foundation drawn to a scale of 1:200 (8 m depth) superimposed. For 4 m depth: scale 40:4000 or 1:100.

Drawing the foundation to 1:100 scale and superimposing on the Newmark chart with point P at the centre.

Estimated influence areas covered by raft $A = 26.25$
Estimated influence areas covered by raft $B = 38.5$
∴   Pressure at depth 4 m below point P

$$= 150 \times \frac{26.25}{200} + 200 \times \frac{38.5}{200} = \underline{\underline{58.1 \text{ kN/m}^2}}$$

For 8 m depth: scale 40:8000 or 1:200.

The foundation drawn to 1:200 scale is shown, superimposed on the Newmark chart with point P at the centre, in Fig. 68(b)

Estimated influence areas covered by raft $A = 22.75$
Estimated influence areas covered by raft $B = 21.5$
∴   Pressure at depth 9 m below point P

$$= 150 \times \frac{22.75}{200} + 200 \times \frac{21.5}{200} = \underline{\underline{38.5 \text{ kN/m}^2}}$$

For 12 m depth: scale 40:1200 or 1:300
Estimated influence areas covered by raft $A = 15$
Estimated influence areas covered by raft $B = 12.5$
∴   Pressure at depth 12 m below point P

$$= 150 \times \frac{15}{200} + 200 \times \frac{12.5}{200} = \underline{\underline{23.8 \text{ kN/m}^2}}$$

For 20 m depth: scale 40:2000 or 1:500
Estimated influence areas covered by raft $A = 7.25$
Estimated influence areas covered by raft $B = 4.25$
∴   Pressure at depth 12 m below point P

$$= 150 \times \frac{7.25}{200} + 200 \times \frac{4.25}{200} = \underline{\underline{9.7 \text{ kN/m}^2}}$$

Check using method of distribution beneath the corner of a rectangle.

Fig. 68(c) shows the method of dividing the foundation into five rectangular areas each with a corner at point P.

Some work on tabulation can be saved by noting that area 1 and area 4 are the same dimensions but area 1 carries a positive load of 150 kN/m² and area 4 carries a negative load of 150 kN/m², i.e. they will cancel one another out and may be ignored.

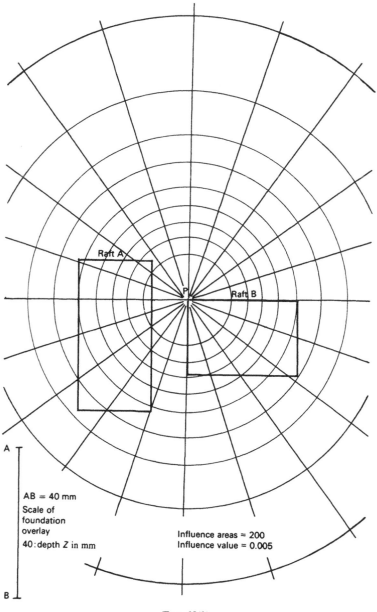

Raft A

P

Raft B

A

AB = 40 mm
Scale of
foundation
overlay
40: depth Z in mm

Influence areas ≈ 200
Influence value = 0.005

B

FIG. 68(b)

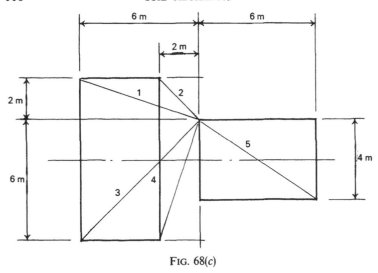

FIG. 68(c)

Depth (metres)	Area	Dimensions (metres)	Load (kN/m²)	m = L/z	n = B/z	k	σᵥ (kN/m²)	For whole area Σσᵥ (kN/m²)
	2	2 × 2	−150	0·5	0·5	0·083	−12·5	
4	3	6 × 6	150	1·5	1·5	0·213	31·95	57·9
	5	6 × 4	200	1·5	1·0	0·192	38·4	
	2	2 × 2	−150	0·25	0·25	0·025	− 3·75	
8	3	6 × 6	150	0·75	0·75	0·137	20·55	38·0
	5	6 × 4	200	0·75	0·5	0·106	21·2	
	2	2 × 2	−150	0·17	0·17	0·015	− 2·25	
12	3	6 × 6	150	0·5	0·5	0·093	13·95	23·7
	5	6 × 4	200	0·5	0·33	0·060	12·0	
	2	2 × 2	−150	0·1	0·1	0·004	− 0·6	
20	3	6 × 6	150	0·3	0·3	0·038	5·7	10·5
	5	6 × 4	200	0·3	0·2	0·027	5·4	

The distribution of pressure below point P by each method, is tabulated below:

Depth (m)	Vertical pressure beneath point P (kN/m²)	
	Newmark	Corner of rectangle
4	58·1	57·9
8	38·5	38·0
12	23·8	23·7
20	9·7	10·5

The slight discrepancy between the values of pressure found by the different methods is due to the difficulty of obtaining accurate estimates of the areas covered on the Newmark chart.

QUESTIONS

1. A point load of 1000 kN acts at the surface of a deep layer of clay. Determine the vertical stress in horizontal layers spaced at 1 metre increments of depth to 8 metres over a range 4 metres either side of the point load, also at 1 metre spacing. From these results plot:

(a) the vertical pressure bulb for 10 kN/m², 20 kN/m² and 40 kN/m².
(b) the distribution of stress directly beneath the load.
(c) the distribution of stress on a horizontal plane at 4 metres depth.

2. A flexible foundation pad 24 m × 12 m carries a uniform load (including own weight) of 120 kN/m². Determine the vertical stress at depths of 3, 6, 12 and 24 metres depths under points:

(a) beneath the centre of the pad.
(b) beneath the centre of a 24 metre edge.
(c) beneath the centre of a 12 metre edge.
(d) beneath a corner.

3. A rectangular building site, $ABCD$ had dimensions $AD = BC = 6$ m and $AB = DC = 8$ m. A building is to be constructed on a flexible foundation to give a contact pressure of 250 kN/m². The foundation covers an L-shaped area along the whole of sides DC and CB and half-way along sides BA and DA (see Fig. 69), the remaining part of the site not being loaded.

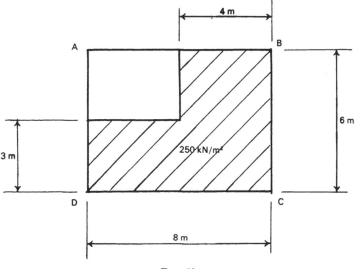

FIG. 69

Use Newmark's chart to determine the vertical pressure at a depth 5 metres below point A.

4. Plot graphically the distribution of vertical pressure at 10 m, 15 m, 20 m and 25 m depths below the corner of a rectangular raft foundation. The raft is 20 m × 10 m in plan and causes a uniform bearing pressure of 120 kN/m² at its underside.

Compare these values graphically with the distribution of vertical pressure beneath the centre of the raft.

5. It is proposed to store material in two paved area A and B as shown in Fig. 70. The pressure exerted by area A is 150 kN/m² and by area B is 300 kN/m².

Calculate the vertical pressures at a depth of 5 m below points 1 and 2.

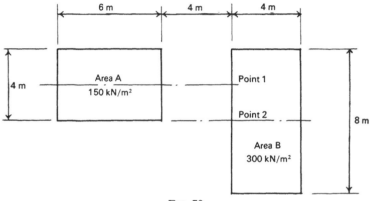

FIG. 70

6. The centres of two columns for a framed building, A and B are 3 metres apart. Column A is supported by a square footing 1·25 m by 1·25 m, the base of which is 2·5 m below ground level. The footing of column B is 1 m by 1 m and its base 1·5 m below ground level. The contact pressure on the clay soil under each footing is 400 kN/m².

Consider the bases as point loads to find the increase in pressure at a depth 5 metres below ground level:

(a) vertically below the centres of the columns.
(b) midway between the columns.

If the coefficient of compressibility of the clay is $m_v = 0·3 \times 10^{-2}$ m²/kN, estimate the differential settlement between the columns if the clay is 5 m thick below column A and 7 m thick below column B.

CHAPTER 8

SOIL PRESSURES ON RETAINING WALLS

ACTIVE PRESSURE

An earth-retaining wall in equilibrium resists horizontal pressure. This pressure could be evaluated by the theory of elasticity, but more practically an empirical coefficient of earth pressure at rest, k_0, is used; i.e. if the weight of soil above any depth z is γz, then the horizontal earth pressure at rest $= k_0 \gamma z$.

Values of k_0:

Normally consolidated clay: 0·5–0·75
Over-consolidated clay: 1·0
Sand (loose–dense): 0·4–0·60

However, in practice, many retaining walls move forwards slightly. When this happens the pressure on the wall is reduced. The minimum value of this pressure *at the point of failure of the soil* is known as the *active pressure*.

Active pressure of cohesionless soil
A dry cohesionless soil is seldom obtained in practice, but the following theories provide a basis for further investigation.

1. *Horizontal soil surface: Rankine's theory*
Rankine's theory, with horizontal soil surface behind a vertical wall, is illustrated in Fig. 71. Consider an element of soil at depth z:

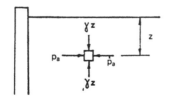

FIG. 71. Soil element: Rankine's theory

Vertical pressure $= \gamma z$
Horizontal pressure $= p_a$ (active pressure at failure)

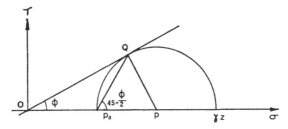

FIG. 72. Mohr circle: Rankine's theory

γz and p_a are the maximum and minimum principal stresses, for which a Mohr circle may be constructed. In Fig. 72:

$$\sin \varphi = \frac{QP}{OP} = \frac{\frac{1}{2}(\gamma z - p_a)}{\frac{1}{2}(\gamma z + p_a)}$$

$$\therefore \quad \gamma z(1 - \sin \varphi) = p_a(1 + \sin \varphi)$$

$$p_a = \frac{1 - \sin \varphi}{1 + \sin \varphi}\gamma z$$

$$p_a = k_a \gamma z$$

where k_a is the coefficient of active earth pressure.

By simple trigonometry it may be shown that:

$$k_a = \frac{1 - \sin \varphi}{1 + \sin \varphi} = \tan^2\left(45 - \frac{\varphi}{2}\right)$$

This theory may also be used to derive an expression for active pressure when the soil surface is not horizontal. However, for this case and the case where the wall is not vertical, it is probably simpler to use the "wedge" theory.

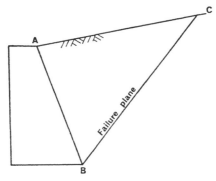

FIG. 73(a). Failure plane: wedge theory

2. Soil surface not horizontal: Coulomb's wedge theory

When the soil behind the wall fails, it will slip along some plane and a wedge of soil will be displaced (Fig. 73(a)). Before failure the wedge of soil above the failure plane is held in equilibrium by three forces (*see* Fig. 73(b)):

W = the weight of the wedge.
P_a = the active force.
R = the reaction of the failure plane.

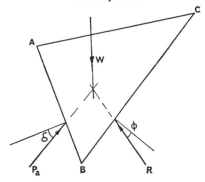

FIG. 73(b). Forces on the wedge

Since the wedge will move downwards on failure, the resultant reaction R is at angle φ (angle of internal friction) to the normal to the failure plane. If the angle of friction between the wall and the soil is δ, then resultant active force P_a is at angle δ to the normal to the wall face.

The value of P_a may be found by resolution of forces, and is dealt with more fully under cohesive soils.

Active pressure of c—φ soils

Bell's solution

Rankine did not consider the question of cohesion in soil, but his theory was developed by Bell to give a simple solution for such soils.

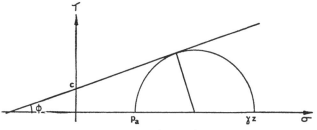

FIG. 74. Mohr envelope

As before, take p_a and γz as the two principal stresses (*see* Fig. 74).
Fig. 74 can be seen to be similar to Fig. 57 for which:

$$\sigma_1 = \sigma_3 N_\varphi + 2c\sqrt{N_\varphi}$$

$\therefore$ for Fig. 74:

$$\gamma z = p_a N_\varphi + 2c\sqrt{N_\varphi} \quad \left[N_\varphi = \tan^2\left(45 + \frac{\varphi}{2}\right) \right]$$

or

$$p_a = \gamma z \frac{1}{N_\varphi} - 2c\frac{\sqrt{N_\varphi}}{N_\varphi}$$

but

$$\frac{1}{\tan^2\left(45 + \dfrac{\varphi}{2}\right)} = \tan^2\left(45 - \frac{\varphi}{2}\right)$$

or

$$\frac{1}{N_\varphi} = k_a$$

$$\therefore \quad p_a = k_a \gamma z - 2c\sqrt{k_a}$$

But for non-cohesive soils $p_a = k_a \gamma z$, therefore cohesion has the
effect of reducing the active pressure by an amount $2c\sqrt{k_a}$, i.e. the
soil is more self-supporting.

Tension cracks
In a cohesive soil the active pressure, p_a, will have a positive value as
long as:

$$k_a \gamma z \geqslant 2c\sqrt{k_a}$$
$$z \geqslant \frac{2c}{\gamma}\frac{\sqrt{k_a}}{k_a}$$
$$z \geqslant \frac{2c}{\gamma}\sqrt{N_\varphi}$$

Therefore when z is less than $(2c\sqrt{N_\varphi})/\gamma$ the active pressure will be
negative, i.e. the soil will be in tension. Since soils will not normally
resist tension forces, vertical cracks will appear which will have a
depth of $(2c\sqrt{N_\varphi})/\gamma$. In cohesive soils with limited drainage, φ is
often taken as zero, and in this case the depth of tension cracks is
$2c/\gamma$.
    As with the Rankine theory, Bell's solution is only easily
applicable to vertical wall face and horizontal earth surface, and
takes no account of wall friction or cohesion.

Effects of water

Apart from the effects of water on the cohesive properties of a soil, there will also be a decrease in active pressure below the water table, since the submerged density of the soil is used:

$$p_a = k_a \gamma' z$$

However, the total pressure on the back of the wall will increase owing to the water pressure. Suitable drainage is usually provided at the back of a retaining wall to reduce this hydrostatic head.

After heavy rainfall, tension cracks in the soil may fill with water, and this will also provide a horizontal pressure.

A simple way of solving problems concerned with active pressure in soils is to draw the pressure diagrams on the back of the wall.

EXAMPLE 30

A vertical wall, 9 m high, supports cohesive soil, the surface of which is level with the top of the wall. The density of the soil is 1900 kg/m³, its cohesion is 20 kN/m² and the angle of shearing resistance is 10°.

Find the active thrust on the wall per lineal metre, assuming that the soil is well drained and neglecting frictional and cohesive forces on the back of the wall.

Find also the active thrust if the soil is waterlogged (saturated density = 2000 kg/m³), with the water table at the surface, assuming the soil strength to be unaltered.

SOLUTION

Coefficient of active pressure $k_a = \tan^2 (45 - \tfrac{10}{2}) = 0.839^2 = 0.704$

$$N_\varphi = \frac{1}{k_a} = \tan^2 (45 + \tfrac{10}{2}) = 1.192^2 = 1.421$$

The pressures on the back of the wall for the drained case are shown in Fig. 75(a)

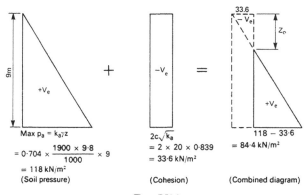

Max $p_a = k_a \gamma z$
$= 0.704 \times \dfrac{1900 \times 9.8}{1000} \times 9$
$= 118 \text{ kN/m}^2$
(Soil pressure)

$2c\sqrt{k_a}$
$= 2 \times 20 \times 0.839$
$= 33.6 \text{ kN/m}^2$
(Cohesion)

$118 - 33.6$
$= 84.4 \text{ kN/m}^2$
(Combined diagram)

FIG. 75(a)

Depth of tension cracks $= z_0$

From similar triangles (Fig. 75(a)):

$$z_0 = \frac{33 \cdot 6}{118 \cdot 7} \times 9 = \underline{\underline{2 \cdot 56 \text{ m}}}$$

Alternative:

$$\text{Depth of tension cracks} = \frac{2c}{\gamma} \sqrt{N_\varphi}$$

$$= \frac{2 \times 20}{1900 \times 9 \cdot 8 / 1000} \times 1 \cdot 192 = 2 \cdot 6 \text{ m}$$

Total thrust $P_a = \frac{1}{2} \times 84 \cdot 4 \times (9 - 2 \cdot 6) = 270$ kN/m run (negative part of combined diagram is ignored).

If the soil is saturated it will also be submerged, with $\gamma' = 2000 - 1000 = 1000$ kg/m^3. The pressure diagrams for this case are shown in Fig. 75(b). *Note:* The water pressure is not taken into account for the determination of depth of tension cracks.

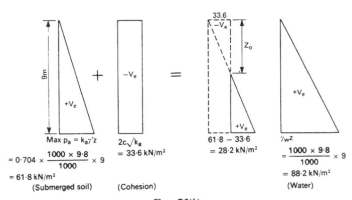

FIG. 75(b)

$$z_0 = \frac{33 \cdot 6}{62} \times 9 = \underline{\underline{4 \cdot 9 \text{ m}}} \qquad \text{(Fig. 75(b))}$$

Total thrust $= \frac{1}{2} \times 28 \cdot 2 \times 5 \cdot 1 + \frac{1}{2} \times 88 \cdot 2 \times 9$

$= \underline{\underline{468 \cdot 8 \text{ kN/m run.}}}$

EXAMPLE 31

Figure 76(a) shows the backfill behind a smooth vertical retaining wall.

(a) Determine the shear force in kN which must be mobilised beneath the base of the wall to prevent movement away from the backfill.

(b) At what height above the base does the total horizontal thrust act?

(c) What would be the total pressure behind the wall if drainage is provided to lower the water table to the base of the wall?

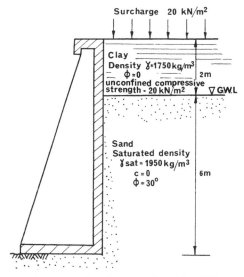

FIG. 76(a). Retaining wall and backfill

The unconfined compressive strength of the clay is given as $20\,\text{kN/m}^2$. Referring to Chapter 6 (Fig. 55), it can be seen that the cohesion therefore must be $10\,\text{kN/m}^2$.

For clay $k_a = 1$
For sand $k_a = \tan^2\left(45 - \frac{30}{2}\right) = \frac{1}{3}$

(a) Total shear force mobilised must be as great as the pressure on the back of the wall.

Effect of surcharge:

$$\text{In clay} = 20\,\text{kN/m}^2$$
$$\text{In sand} = 20 \times \tfrac{1}{3} = 6\cdot67\,\text{kN/m}^2$$

Effect of clay:

$$\text{Active pressure at base of clay} = \frac{1750 \times 2 \times 9\cdot8}{1000}$$
$$= 34\cdot3\,\text{kN/m}^2$$
$$\text{Active pressure at base of sand} = 34\cdot3 \times \tfrac{1}{3} = 11\cdot43\,\text{kN/m}^2$$
$$\text{Cohesion} = 2c\sqrt{k_a} = 2 \times 10 \times 1$$
$$= 20\,\text{kN/m}^2$$

Effect of submerged sand:

$$\text{Active pressure at base of sand} = \frac{950 \times 6 \times 9\cdot8}{1000} \times \frac{1}{3}$$
$$= 18\cdot6\,\text{kN/m}^2$$

Effect of water:

  Maximum pressure due to water $= 9\cdot8 \times 6 = 58\cdot8$ kN/m^2

  These values are shown in Fig. 76(b), together with the lever arm of each pressure diagram about the base.

Note: The surcharge exactly cancels out the cohesion and therefore no tension cracks will form.

Total pressure $= 20 \times 2 + 6\cdot67 \times 6 + \frac{1}{2} \times 34\cdot3 \times 2 + 11\cdot43 \times 6 -$
$\qquad\qquad\quad 20 \times 2 + \frac{1}{2} \times 18\cdot6 \times 6 + \frac{1}{2} \times 58\cdot8 \times 6 = \underline{\underline{375\ \text{kN}}}$

  (b) Moments about base (let h = height of total thrust)

$375\,h = 6 \times 6\cdot67 \times 3 + \frac{1}{2} \times 34\cdot3 \times 2 \times 6\cdot67 + 6 \times 11\cdot43 \times 3$
$\qquad\quad + \frac{1}{2} \times 18\cdot6 \times 6 \times 2 + \frac{1}{2} \times 58\cdot8 \times 6 \times 2$
$\underline{\underline{h = 2\cdot72\ \text{m}}}$

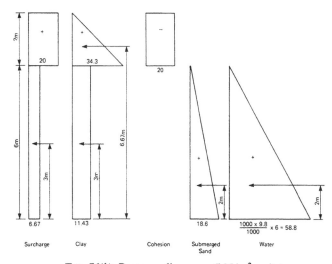

FIG. 76(b). Pressure diagrams. (kN/m^2 units)

  (c) The net effect of draining the soil would be:

  (i) The water pressure ($\frac{1}{2} \times 58\cdot8 \times 6 = 176\cdot4$ kN) would not be taken into account.

  (ii) The sand would no longer be submerged and the submerged pressure of ($\frac{1}{2} \times 18\cdot6 \times 6 = 55\cdot8$ kN) would be increased to:

$$\frac{1}{2} \times \frac{1}{3} \times \frac{1950 \times 9\cdot8}{1000} \times 6^2 = 114\cdot7\ \text{kN}$$

Therefore with drainage:

$$\text{Total pressure} = 375\cdot1 - 176\cdot4 - 55\cdot8 + 114\cdot7$$
$$= \underline{\underline{257\cdot6\ \text{kN}}}$$

Drainage clearly gives lower values of total pressure and should therefore always be considered when designing a retaining wall.

Coulomb's wedge theory
The wedge theory applied to cohesive soils is similar to that for cohesionless soils, except that two further forces have to be considered acting on the wedge:

(a) Cohesion along the failure plane.
(b) Cohesion along the plane of the wall.

Also, for the depth of the tension cracks no friction or cohesion can develop.

EXAMPLE 32
Figure 77(a) shows a retaining wall and silt backfill. If the bulk density of the silt is 1800 kg/m^2, the cohesion 20 kN/m^2, the wall adhesion 12 kN/m^2, the angle of shearing resistance of the retained earth 19° and the angle of friction between the wall and the retained earth is 14°:

(a) Find the depth of tension cracks z_0.
(b) Find the maximum active thrust on the wall by the method of wedges using the trial slip planes shown.

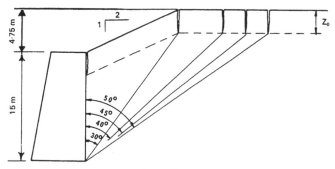

FIG. 77(a)

SOLUTION

(a)   Depth of tension cracks $z_0 = \dfrac{2c}{\gamma} \times \sqrt{N_\varphi}$

$$= \dfrac{2 \times 20}{1800 \times \dfrac{9 \cdot 8}{1000}} \tan\left(45 + \tfrac{19}{2}\right)$$

$$= \underline{\underline{3 \cdot 2 \text{ m}}}$$

(b) The forces acting on each wedge are as shown in Fig. 77(b) and their values given in the table.

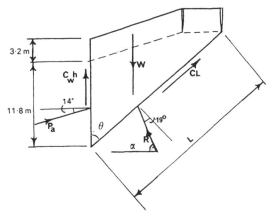

FIG. 77(b). Forces on wedge

Angle of trial slip plane with vertical (θ)	Area of wedge (m^2)	Weight of wedge (kN)	Length of slip plane (m)	Cohesion (kN)	Angle of reaction R with horizontal (α)
50°	203·3	3586	25·8	516	69°
45°	166·9	2944	23·4	468	64°
40°	136·3	2404	21·6	432	59°
30°	86·5	1526	19·1	382	49°

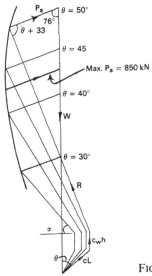

FIG. 77(c)

For each wedge:

Wall adhesion acts over 11·8 m

∴ Wall adhesion = 11·8 × 12 = 141·6 kN

Angle of P_a with vertical = 76°

These forces are plotted on superimposed polygons of forces in Fig. 77(c). From the figure the maximum value of P_a is 850 kN.

PASSIVE RESISTANCE

In certain cases a wall may be pushed towards the soil, as in bridge abutments on expansion or in front of a retaining wall. The maximum value of this pressure at the point of failure of the soil is known as the *passive pressure* or *resistance*.

Passive resistance of a cohesionless soil

Horizontal soil surface: Rankine's theory

For a vertical wall with horizontal soil surface the passive pressure p_p will be the maximum principal stress and γz the minimum:

$$\therefore \quad p_p = \frac{1 + \sin \varphi}{1 - \sin \varphi} \gamma z$$

$$= k_p \gamma z$$

where k_p is the coefficient of passive earth pressure

$$= \tan^2 \left(45 + \frac{\varphi}{2} \right).$$

Note: k_p is numerically the same as N_φ, but it is convenient to use the two symbols.

Soil surface not horizontal: Coulomb's wedge theory

When a soil is subjected to passive pressure the wedge is forced upwards, and therefore P_p and R will act as shown in Fig. 78. This can only be applied when $\delta < \dfrac{\varphi}{3}$, as beyond that figure the slip surface is not a flat plane.

Passive resistance of c—φ soil

Bell's solution

$$\sigma_1 = \sigma_3 N_\varphi + 2c\sqrt{N_\varphi}$$

$$\therefore \quad p_p = \gamma z N_\varphi + 2c\sqrt{N_\varphi}$$

or $\qquad p_p = k_p \gamma z + 2c\sqrt{k_p}$

i.e. cohesion increases the value of passive thrust.

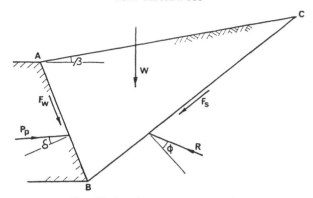

FIG. 78. Passive pressure on wedge

SHEET-PILE WALLS

Sheet-pile walls generally depend upon the passive resistance of the soil to maintain equilibrium. The problem when sheet piling is to decide the depth piles should be driven for stability and economy. If there is a considerable depth of soil to be retained the sheet piles will normally need to be tied back near the surface. For small depths the sheet pile may act as a cantilever.

Fixed earth support
Cantilever sheet-pile walls. When the sheet piles are subjected to active pressure from the retained soil the soil in front of the piles is subjected to passive pressure. These active and passive pressures are shown in Fig. 79(a). The reaction at the foot of the pile R_p would in

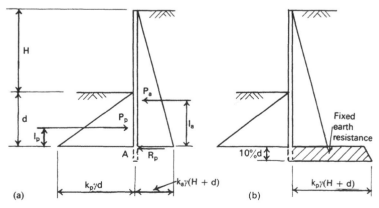

FIG. 79(a). Force diagram.   (b). Fixed earth reaction fact be made up of a zone of passive pressure behind the wall as shown in Fig. 79(b), the point of rotation of the piles being a short distance up from the base. The assumption of this reaction at the base is known as the fixed earth support theory. Taking moments about A:

$$P_p \times l_p = P_a \times l_a$$

From this equation, d may be found. The depth of driving is taken 10 per cent greater than d to allow for the fixed earth reaction.

The calculation of d involves solving a cubic equation and it will be found in cohesionless soils that the depth of driving is greater than the height of soil supported. Cantilever sheet piles are therefore uneconomical for such soils. In cohesive soils the economic height of supported soil depends on the value of cohesion and calculated values should be treated with *extreme caution*. For these reasons no calculation for cantilever sheet piles will be attempted in this volume.

Anchored bulkhead: fixed earth support. If the retained soil is high, a tie rod will be required to maintain equilibrium. The reaction R_p may again be assumed to exist and the fixed earth support theory is used (*see* Fig. 80). However, for anchored sheet piles the free earth support theory gives a simpler calculation.

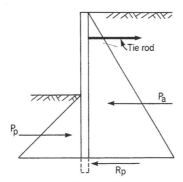

FIG. 80. Anchored bulkhead (fixed earth support)

Free earth support

Anchored bulkhead: Another way of considering the stability of sheet piles is to assume that they rotate about the base of the pile. In this case there will be no reaction at A and the sheet pile is supported by the passive pressure in front of the face and the tie rod (*see* Fig. 81). This case is easily solved by taking moments about the tie rod and has also been shown by experiment to give a more correct result.

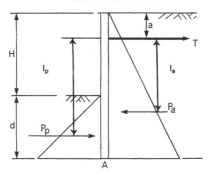

FIG. 81. Anchored bulkhead (free earth support)

EXAMPLE 33

An anchored sheet-pile wall, shown in Fig. 82, retains soil of height 6 m, the piles having a total length of 9·75 m. The soil has a bulk density of 1900 kg/m³, a negligible cohesion and an angle of shearing resistance of 30°. The anchor tie rods are at 1·25 m below the surface of the soil.

(a) Find the active thrust on the piling in kN per horizontal metre.

(b) Find the maximum possible passive pressure.

(c) Using the "Free Earth Support Theory" determine the passive press-ure mobilised and hence the factor of safety of the sheet pile wall.

(d) Find the tension in the tie rods if they are spaced at 5 m centres.

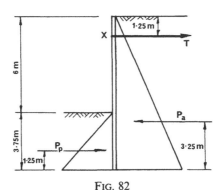

FIG. 82

SOLUTION

(a) $k_a = \tan^2 (45 - \frac{30}{2}) = \frac{1}{3}$ $\qquad\qquad\qquad k_p = \tan^2 (45 + \frac{30}{2}) = 3$

$P_a = \frac{1}{2} \times \frac{1}{3} \times (1·9 \times 9·8)9·75^2 = \underline{295 \text{ kN}}/\text{horizontal metre}$

(b) $P_p = \frac{1}{2} \times 3 \times (1·9 \times 9·8) \times 3·75^2 = \underline{\underline{393 \text{ kN}}}$ (maximum possible)

(c) To find the passive resistance mobilised take moments about the tie rod at X

$$7{\cdot}25\,P_p = 5{\cdot}25 \times 295$$
$$P_p = 214\,\text{kN (mobilised)}$$
$$\text{Factor of safety} = \frac{393}{214} = \underline{\underline{1{\cdot}8}}$$

(d) Tension in the rods = $(295 - 214)5$
$$= \underline{\underline{405\,\text{kN}}}$$

QUESTIONS

1. A vertical retaining wall supports a cohesive backfill having a cohesion of 24 kN/m² and an angle of shearing resistance of 8° (*see* Fig. 83). The height of the wall is 12 m and G.W.L. is 4 m below the crest. The bulk density of the soil above G.W.L. is 1640 kg/m³· and the saturated soil density is 2000 kg/m³. The ground surface behind the wall, which is horizontal and level with the wall crest, carries a uniform surcharge of 12 kN/m².

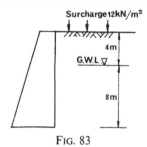

Surcharge 12kN/m²

4 m

G.W.L ▽

8 m

FIG. 83

(*a*) Determine the total horizontal thrust on the wall in kN per lin. metre if the angle of wall friction is zero and tension cracks have formed.

(*b*) To what depth is the pressure on the wall zero?

(*c*) At what height above the base does the total horizontal thrust act?

2. A retaining wall 5 m high supports a backfill consisting of 2 m of sandy clay overlying 3 m of sand. G.W.L. coincides with the upper surface of the sand. The soil constants are as follows. *Sandy clay:* density = 1840 kg/m³, cohesion = 11·5 kN/m²; $\varphi = 10°$. *Sand:* density = 1930 kg/m³; $\varphi = 35°$, $c = 0$.

(*a*) Determine the total active thrust on the retaining wall and its point of action, assuming tension cracks have developed.

(*b*) If it were possible to lower G.W.L. by 2 m, without altering the soil properties, what then would be the active thrust on the wall? Again assume tension cracks will develop.

3. A retaining wall, of height 9·75 m, has the earth face battered at 8 vertical to 1 horizontal. The backfill, which is level with the top of the wall, is clay with a density of 1920 kg/m³, cohesion 20 kN/m² and $\varphi = 0$. A trial slip plane is chosen, making 35° with the horizontal.

Find, for this slip plane, the thrust on the wall per lineal metre. Allow for tension cracks, and assume the water table to be below the base of the wall. Assume wall adhesion equals 0·5 × cohesion of the soil.

Determine analytically or graphically the value of the maximum active thrust on the wall. The density of the sand is 1770 kg/m³.

(b) If the sand were replaced exactly by a cohesive frictional soil in which tension cracks had appeared, explain, with suitable diagrams, how the value of the maximum active thrust may be determined graphically if the adhesion between the wall and soil were considered.

4. A retaining wall slopes up and away from a cohesionless backfill with a batter of 1 in 10. Density of backfill = 1850 kg/m³. Angle of surcharge = 5°. Angle of internal friction = 35°. Angle of wall friction = 25°. Vertical height of wall = 10·67 m.

Determine by graphical construction the maximum active thrust on the wall.

5. Figure 84 shows an anchored sheet-pile bulkhead with "free earth support". Density of cohesionless soil = 1850 kg/m³. Angle of shearing resistance = 35°. The bulkhead is securely anchored by horizontal tie rods 1 m below the top. Calculate the minimum passive thrust that must be mobilised on the embedded length of the bulkhead for equilibrium. Is the depth of penetration of the sheet piles sufficient to develop this thrust?

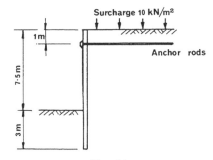

FIG. 84

CHAPTER 9

STABILITY OF SLOPES

A fairly common engineering failure is slipping of an embankment or cutting, and considerable research has been carried out into the causes of such failures. Water is frequently the cause of earth slips, either by eroding a sand stratum, lubricating a shale or increasing the moisture content of a clay, and hence decreasing the shear strength. When a slip in a clay soil occurs it is frequently found to be along a circular arc, and therefore this shape is assumed when studying the stability of a slope. This circular arc may cut the face of the slope, pass through the toe or be deep-seated and cause heave at the base (*see* Fig. 85). Prediction of the most likely failure plane relies heavily on experience, based on the study of past cases.

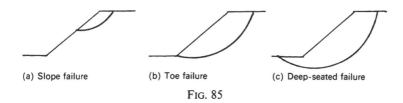

(a) Slope failure   (b) Toe failure   (c) Deep-seated failure

FIG. 85

The cause of slope failure in a cutting will be quite different from that in an embankment. A cutting is an *unloading* case where soil is removed, hence causing relief of stress in the soil. The soil resistance dissipates with time and a part of the engineer's problem will be to predict the soil properties during the design life of a cutting (*see* residual shear, Chapter 6).

Embankments and spoil heaps, on the other hand, are *loading* cases and the construction period is the most critical period, owing to the build up of pore pressures during construction, with the consequent reduction in effective stress (*see* pore-pressure parameters, Chapter 6). In time, these excess pore pressures dissipate and the shear resistance of the embankment increases, although consolidation may now become the major problem.

In both cases the study of the variation of pore-water pressure within the soil mass is of paramount importance, since only in this way can reasonable values of the parameters c and φ be determined. Water seepage will set up seepage pressures (*see* Chapter 4) which may induce failure of the slope. Such water movement frequently occurs in the vicinity of major earthworks such as an earth dam or cuttings below the natural water table. Seepage pressures can be monitored with piezometers on site and it may be necessary to provide suitable drainage to control the flow of water.

Detailed discussion of the factors involved in predicting the most likely slope failure plane is beyond the scope of this book. This chapter is therefore confined to the *analysis* of slope failure, assuming previously selected values of the parameters c and φ. This analysis would itself form a part of the decision as to the most likely failure plane.

Tension cracks

In any cutting, tension cracks may form at the top of the slope, and these cracks may be the first indication of slope failure. In Chapter 8 it was shown that the theoretical depth of these tension cracks, Z_0 $= 2c \sqrt{N_\varphi}/\gamma$.

If these tension cracks fill with water, the hydrostatic pressure will reduce the stability of the slope. The problem is greater if the water freezes in the cracks. Such cracks along the top of a slope therefore, should be noted with care and dealt with at the earliest opportunity.

Vertical cuts

If a vertical cut is made in a clay soil it will remain stable for a short period, largely due to negative pore pressure caused by removal of the load. It must be stressed that *it is a very dangerous practice to leave a vertical cut unsupported at any time*. However, it may be useful to predict how deep such a cut may be made.

It can be shown that the critical height of a vertical cut, H_c $= 2.67c/\gamma$. This assumes $\varphi = 0$ as would be the case for such a short-term case.

$\varphi = 0$ CONDITION

Since vertical cuts should always be supported either by suitable timbering or sheet piles, steep-sided temporary cuts are preferred for underground work which is to be backfilled upon completion. This may be considered a short-term case (depending on the time of exposure) and the $\varphi = 0$ condition assumed.

A simple undrained shear test can be used to obtain the value of apparent cohesion.

Consider a likely circular slip surface BC (*see* Fig. 86(a)) with centre at O

The disturbing moment of the cylinder of soil about O
$$= \text{Weight of soil} \times \text{Distance } d = \text{Wd}$$

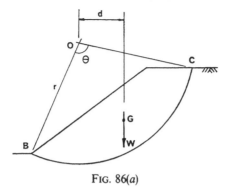

FIG. 86(a)

The resisting moment preventing the soil from moving is all due to cohesion along BC, which has a lever arm equal to radius r about O,

i.e.   Resisting moment = Cohesion × Length BC × Radius r
$$= c \times r\theta \times r$$
$$= cr^2\theta$$

Factor of safety against slip

$$= \frac{\text{Resisting moment}}{\text{Disturbing moment}} = \frac{cr^2\theta}{Wd}$$

A series of slip circles are checked, and the one with the lowest factor of safety is the critical slip surface.

Frequently a tension crack appears at some distance from the top of an earth slope and parallel to it. The bottom of this crack may be taken at a depth of $2c/\gamma$ and is a point through which the slip circle will pass. No cohesive resistances can develop to this depth, and therefore the surface resisting rotation is BC' (see Fig. 86(b)). This crack may fill with water and exert hydrostatic pressure, which will also have a moment about O.

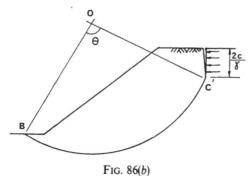

FIG. 86(b)

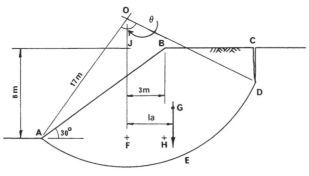

FIG. 87

EXAMPLE 34

A temporary cutting has the profile shown in Fig. 87. The material is homogeneous clay of density 1800 kg/m³, cohesion 50 kN/m² and angle of shearing resistance zero.

Find the factor of safety for the slip circle shown, allowing for a tension crack which may be filled with water.

SOLUTION

Depth of tension cracks, $Z_0 = \dfrac{2 \times 50}{1 \cdot 8 \times 9 \cdot 8} = 5 \cdot 7$ m

Draw the figure to scale and calculate the area $ABCDE$, the position of the centroid G and the angle θ. The values for this slip area are:

$$\text{Area } ABCDE = 184 \text{ m}^2$$
$$\text{Lever arm } la = 4 \cdot 6 \text{ m}$$
$$\text{Angle } \theta \quad = 89°$$

Disturbing moment $= 184 \times 1 \cdot 8 \times 9 \cdot 8 \times 4 \cdot 6 = 14\,930$ kNm

Resisting moment $= AED \times 50 \times 17$

$$= \left(17 \times 89 \times \frac{\pi}{180}\right) \times 50 \times 17 = 22\,446 \text{ kNm}$$

Assuming water in tension cracks

$$AH = 8\sqrt{3} = 13 \cdot 8 \text{ m}$$
$$\therefore \quad AF = 10 \cdot 8 \text{ m}$$
$$OF = \sqrt{17^2 - 10 \cdot 8^2} = 13 \cdot 1 \text{ m}$$
$$OJ = 5 \cdot 1 \text{ m}$$

Moment from water in crack $= \frac{1}{2} \times 9 \cdot 8 \times 5 \cdot 7^2 \times (5 \cdot 1 + \frac{2}{3} \times 5 \cdot 7)$
$$= 1416 \text{ kNm}$$

Disturbing moment $= 14\,930 + 1416 = 16\,346$ kNm

$\therefore$ Factor of safety $= 22\,446/16\,346 = \underline{\underline{1 \cdot 37}}$

If the cutting is to remain open long enough for the soil to drain and a value of φ to develop, or if the soil is not saturated, then a c—φ analysis must be carried out.

c—φ SOILS

For c—φ soils the shear resistance along the slip plane varies with the normal force. If, therefore, the whole or part of the shear strength is due to friction a graphical approximation is used. A possible slip circle is chosen and divided into strips of equal width (*see* Fig. 88 (*a*)). Consider one strip (*see* Fig. 88 (*b*)). Vertical weight W can be considered in two components: (1) $N = W\cos\alpha$ at right angles to arc of slip, (2) $T = W\sin\alpha$ tangential to arc of slip.

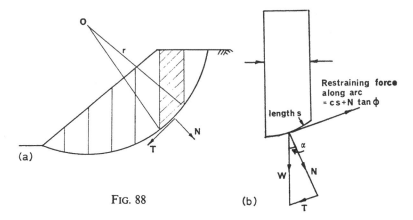

Restraining force
along arc
= c s + N tan ϕ

length s

(a)

FIG. 88   (b)

For one strip disturbing moment about centre $O = T \times r$

For whole area disturbing moment $= r\Sigma(T)$

The restraining force on one strip is made up of cohesion $c \times s$ and frictional force $N \tan \varphi$.

For one strip resisting moment about centre $O = (cs + N \tan \varphi) r$

For whole area resisting moment $= r(cr\theta + \tan \varphi \Sigma N)$

$$\text{Factor of safety} = \frac{cr\theta + \tan \varphi \Sigma N}{\Sigma(T)}$$

c, r, θ and φ are readily determined, and N and T may be found for each strip and summed.

EXAMPLE 35

Figure 89(*a*) shows a cutting that has been made in a homogeneous silty clay. The soil constants for undisturbed samples are $c = 20$ kN/m^2 and $\varphi = 8°$.

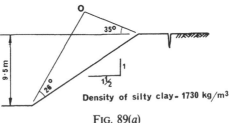

FIG. 89(a)

Allowing for the development of tension cracks, what is the factor of safety associated with a slip circle (centre O) passing through the toe of the bank? Is a toe failure probable?

SOLUTION

Depth of tension cracks $z_0 = \dfrac{2c\sqrt{N\varphi}}{\gamma} = \dfrac{2 \times 20 \tan 49°}{1\cdot73 \times 9\cdot8} = 2\cdot7$ m

Figure 89(b) shows the slip circle divided into 14 strips, each 1·5 m wide. The weight of each strip may be represented by the length of the strip and is drawn vertically underneath. The normal and tangential forces are drawn to complete the force diagram for each strip.

Magnitudes of vectors (in metre units)			
Strip No.	N	+T	−T
1	1·0		0·5
2	2·8		1·0
3	4·4		1·2
4	5·6		0·9
5	7·0		0·3
6	7·7	0·4	
7	8·8	1·2	
8	9·5	2·4	
9	9·5	3·4	
10	8·9	4·4	
11	7·5	4·9	
12	6·1	5·0	
13	4·2	4·6	
14	2·0	3·5	
	85·0	29·8	3·9

Disturbing force $= \Sigma T = (29\cdot8 - 3\cdot9) \times 1\cdot6 \times 1\cdot73 \times 9\cdot8 = 703$ kN

Resisting forces: $cr\theta = 20 \times 16\cdot3 \times \tfrac{93}{180}\pi = 529$ kN

$\Sigma N \tan \varphi = 85 \times 1\cdot6 \times 1\cdot73 \times 9\cdot8 \times 0\cdot1405 = 324$ kN

Factor of safety $= \dfrac{529 + 324}{703} = \underline{\underline{1\cdot21}}$

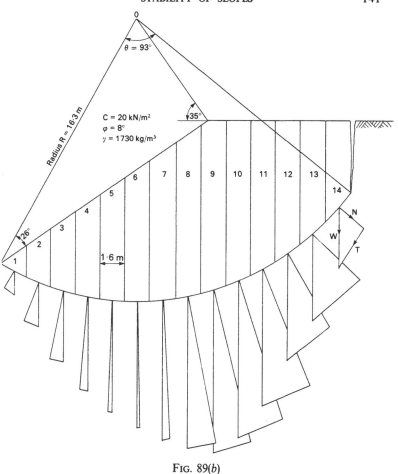

FIG. 89(b)

A toe failure is not imminent, but the factor of safety is low.

EFFECTIVE STRESS ANALYSIS

If the equation for factor of safety were to be expressed in terms of effective stress, the resisting force would become $c'r\theta + \tan \varphi' \Sigma(N - u \times s)$

where: u = the pore water pressure at the bottom of each vertical strip.

s = the length of slip plane for each vertical strip (see Fig. 88(b)).

Also referring back to Fig. 88(b) the two components N and T could have been written $W\cos\alpha$ and $W\sin\alpha$ respectively, where α is as shown in Fig. 88(b). The equation would then be:

$$\text{Factor of safety} = \frac{c'r\theta + \tan\varphi'\Sigma(W\cos\alpha - u \times s)}{\Sigma W\sin\alpha}$$

A pore pressure coefficient, r_u, may be introduced into this equation where,

$$r_u = \frac{\text{pore pressure at any point}}{\text{overburden pressure at that point}} = \frac{u}{\gamma h}$$

h being the height of soil above the point considered and γ the soil density.

Referring to one of the strips shown in Figs. 88(a) and (b).

$$r_u = \frac{u}{\gamma h} = \frac{u \times b}{\gamma h \times b} = \frac{ub}{W}$$

$$\text{also } \frac{b}{s} = \cos\alpha \text{ or } b = s\cos\alpha$$

$$\therefore \quad r_u = \frac{u \times s\cos\alpha}{W}$$

$$\text{or } u \times s = \frac{r_u W}{\cos\alpha}$$

The expression for factor of safety referred to effective stress would now be,

$$\text{Factor of safety} = \frac{c'r\theta + \tan\varphi'\Sigma\left(W\cos\alpha - \dfrac{r_u W}{\cos\alpha}\right)}{\Sigma W\sin\alpha}$$

$$= \frac{c'R\theta + \tan\varphi'\Sigma W(\cos\alpha - r_u\sec\alpha)}{\Sigma W\sin\alpha}$$

A more rigorous analysis could be carried out but the above expression gives a rapid solution with errors which tend towards safety.

The value of the pore-pressure coefficient, r_u, is assumed constant throughout the cross-section.

EXAMPLE 36

Figure 90(a) shows a section through a site in which compacted fill is laid to a depth of 2·9 m and then an embankment 8 m high. The underlying rock is a hard shale.

At the end of construction the properties of the compacted fill are: $\gamma = 1900$ kg/m^3, $c' = 25$ kN/m^2 and $\varphi' = 20°$.

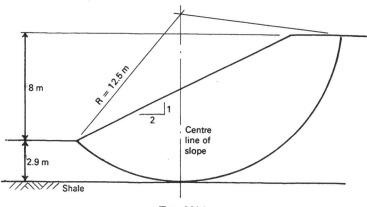

FIG. 90(a)

The pore pressure ratio has an average value of 0·3.
Determine the factor of safety for the slip circle shown.

SOLUTION
The section is divided into vertical strips and the force diagrams constructed as before (*see* Fig. 90(b)).

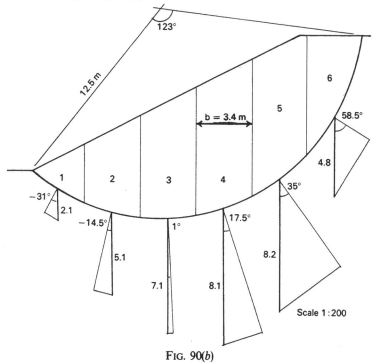

FIG. 90(b)

From Fig. 90(b) the results are tabulated as follows:

Strip	W	α	cos α	sec α	sin α (+)	sin α (−)
1	2·1	−31°	0·857	1·167		−0·515
2	5·1	−14·5°	0·981	1·033		−0·250
3	7·1	1°	1·000	1·000	0·017	
4	8·1	17·5°	0·954	1·048	0·301	
5	8·2	35°	0·819	1·221	0·574	
6	4·8	58·5°	0·523	1·912	0·853	
Σ	35·4		5·121	7·381	1·745	−0·765

Factor of safety =

$$\frac{25 \times 12·5 \times 123 \times \pi/180 + 0·364 \times (35·4 \times 3·4 \times 1·9 \times 9·8)(5·121 - 0·3 \times 7·381)}{(35·4 \times 3·4 \times 1·9 \times 9·8)(1·745 - 0·765)}$$

$$= \underline{\underline{1.39}}$$

STABILISATION OF SLOPES

When designing a slope or attempting to stabilise an existing failure, a number of possible remedies are open to the engineer.

From the preceding discussion, it will be seen that control of water must be considered. Suitably designed drainage should minimise any seepage pressures which may be set up and will also reduce pore-water pressures, thus increasing effective stress, and therefore, the stability of the slope.

In embankment slopes, horizontal layers of coarse material may be included to facilitate drainage and provision must be made to dispose of the water from these layers.

In cuttings, surface drainage will prevent softening of the upper layers of soil, but will do little to increase overall stability. Installation of deep-seated drainage in a cutting can be very expensive and some method of loading, or unloading, the slope may provide a better solution.

For natural hillslopes, the slip surface is generally along a plane parallel to the ground surface and at a fairly shallow depth. In this situation, surface drainage, provided it reaches beyond the failure plane, may well be sufficient.

Each case has its own inherent problems, and the following examples are merely an indication of the types of problem and their corresponding solutions.

EXAMPLE 37
Details of a temporary excavation are shown in Fig. 91(a). The cutting is showing signs of failure along the slip surface indicated and remedial

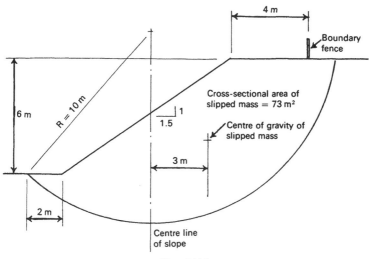

FIG. 91(a)

measures are urgently required. The site is limited by the boundary fence shown, and at least 2 m clearance is required inside this fence. Material may be stored at the base of the cutting, and will exert a uniform loading of 50 kN/m².

Assuming the $\varphi = 0$ condition applies and no tension cracks have formed, suggest suitable remedial measures. The bulk density of the soil is 1750 kg/m³.

SOLUTION
Since the cutting is showing signs of instability, it is reasonable to assume the factor of safety is 1.

Therefore by "back analysis" (see Fig. 91(b) for angle θ)

$$1 = \frac{c \times 10^2 \times 121 \times \pi/180}{1.75 \times 9.8 \times 73 \times 3}$$

$$c = 17.8 \text{ kN/m}^2$$

i.e. Resisting moment $= 17.8 \times 10^2 \times 121 \times \pi/180 = 3759 \text{ kNm}.$

Disturbing moment $= 1.75 \times 9.8 \times 73 \times 3 = 3756 \text{ kNm}.$

An immediate remedial measure would be to add the loading at the toe (see Fig. 91(b)).

Counter balance moment $= 2 \times 50 \times 5.5 = 550 \text{ kNm}$

$$\text{Factor of Safety} = \frac{3756 + 550}{3756} = 1.15$$

Further stabilisation could be obtained by cutting the slope back to the minimum angle with the horizontal as shown in Fig. 91(b). However, a better solution would be to cut a 2 m wide berm in the slope as shown in Fig. 91(c).

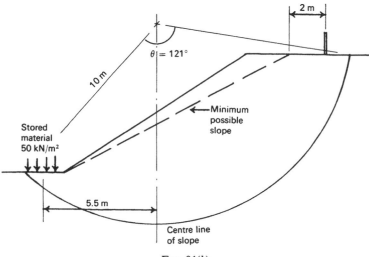

FIG. 91(b)

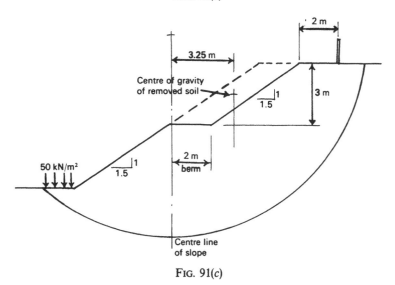

FIG. 91(c)

This would reduce the disturbing moment by removing soil from the right-hand side of the centre of the slip circle only, i.e.:

Reduction in disturbing moment $= 1 \cdot 75 \times 9 \cdot 8 \times 3 \times 2 \times 3 \cdot 25$
$$= 334 \text{ kNm.}$$

Factor of Safety $= \dfrac{3756 + 550}{3756 - 334} = \dfrac{4306}{3422} = 1 \cdot 25$

If the loading were later removed

$$\text{Factor of Safety} = \frac{3756}{3756 - 334} = 1\cdot1$$

This would appear to give a reasonable solution for a temporary contract of this type providing the toe loading is not removed until the excavation is to be backfilled.

EXAMPLE 38
Figure 92(a) gives details of an existing canal bank. The soil properties are $\gamma = 1820 \text{ kg/m}^3$, $c = 7 \text{ kN/m}^2$ $\varphi = 20°$ and no tension cracks have formed. Check the stability of the bank along the slip surface shown:

(a) when the canal is full
(b) if the canal were rapidly drained.

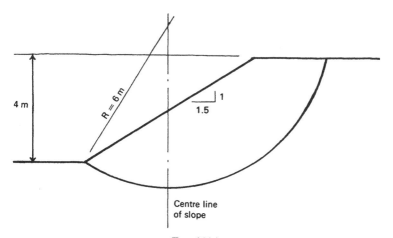

FIG. 92(a)

SOLUTION
Figure 92(b) shows the force diagrams for five vertical strips. The values of ordinate W and angle α are set out in the table.

Strip	W	α	$W\cos\alpha$	$W\sin\alpha$	h_w	$u \times s$
1	1·0	−21°	30	−11·5	0·6	11·3
2	2·6	−4°	83	−5·8	1·8	31·8
3	3·7	14°	115	28·8	3·0	54·6
4	3·7	33°	100	64·8	3·7	77·8
5	2·0	58°	34	55·1	2·0	66·6
Σ			362	131·4		242·1

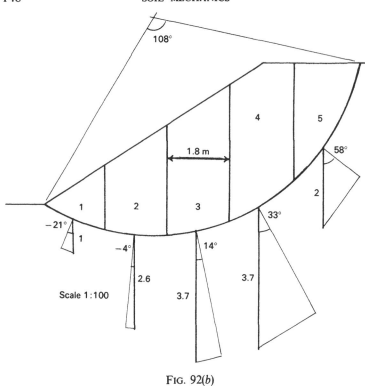

FIG. 92(b)

Note: The columns $W \cos \alpha$ and $W \sin \alpha$ have been multiplied by $1·82 \times 9·8 \times 1·8$ to give kN units.

(*a*) Canal full:

$$\text{Factor of Safety} = \frac{7 \times 6 \times 108 \times \pi/180 + 362 \times 0·364}{131·4}$$

$$= \frac{79 + 132}{131·4} = \underline{\underline{1·6}}$$

(*b*) If the water were rapidly lowered the disturbing moment would not immediately change since the soil in the bank would not drain. Excess pore-water pressures would however be induced which, for each strip, would be equivalent to the height of the excess water above the failure surface, h_w.

From the table, column $u \times s = h_w \times 9·8 \times 1·8 \sec \alpha$

Water rapidly lowered:

$$\text{Factor of Safety} = \frac{79 + 0·364\,(362 - 242·1)}{131·4} = \underline{\underline{0·93}}$$

It can be seen that rapid lowering of the water table would be likely to cause failure of the canal bank. Lowering should, therefore, be carried out at a rate that allows the excess pore pressures in the bank to dissipate.

EXAMPLE 39

An embankment of compacted fill for a road has been designed, and the initial analysis gave the following results at completion of construction:

cohesive force on slip plane $= 1200$ kN; weight of most dangerous slip section $= 4200$ kN; $\tan \varphi' = 0.4$; $\Sigma \cos \alpha = 6$; $\Sigma \sec \alpha = 8$; $\Sigma \sin \alpha = 1$; Pore-pressure coefficient, $r_u = 0.5$.

Estimate the factor of safety and comment on the result.

If a factor of safety of 1·6 is required, without alteration to the geometry of the embankment, suggest a suitable solution and show how the required factor of safety could be obtained.

SOLUTION

$$\text{Initial factor of safety} = \frac{1200 + 0.4 \times 4200\,(6 - 0.5 \times 8)}{4200}$$

$$= \underline{1.08}$$

This would be an unsafe construction.

To improve the factor of safety and maintain the geometry of the embankment, it is necessary to reduce the pore-pressure coefficient. This could be done by slowing the rate of construction, but construction costs would be considerably increased.

An alternative is to provide horizontal drainage blankets at intervals in the embankment, thus allowing dissipation of pore-water pressure. The excess pore-water pressures may be monitored during construction and the value of r_u kept to a maximum permissible value. This would give a factor of safety of

$$1.6 = \frac{1200 + 0.4 \times 4200\,(6 - r_u \times 8)}{4200}$$

$$r_u = 0.34$$

Thus if the pore-pressure coefficient is not allowed to exceed 0·3 the required factor of safety will be achieved.

QUESTIONS

1. Figure 93 shows a cutting that has been made in a homogeneous silty clay. The soil constants for undisturbed samples are $c = 48$ kN/m² and $\varphi = 0°$. The bulk density of the soil is 1800 kg/m³. The figure also shows the centre O of the "most dangerous circle". What is the factor of safety associated with this circle? Allow for tension cracks filled with water.

2. Estimate the factor of safety for the trial circle shown in Fig. 94. Investigation shows that tension cracks extend the full depth of the upper clay stratum. Make the usual allowance for the tension crack being filled

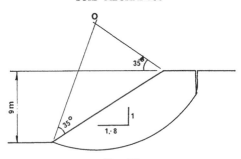

FIG. 93

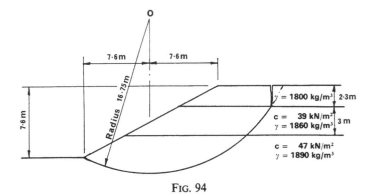

FIG. 94

with water during periods of heavy rain storms. Assume $\varphi = 0°$ throughout.

3. Figure 95 shows the dimensions of a cutting and a trial slip surface. The soil properties are: bulk density, $\gamma = 2100$ kg/m^3, shear parameters, $c = 40$ kN/m^2 $\varphi = 15°$.

By graphical means, estimate the factor of safety for the trial slip surface shown. Assume tension cracks have formed and are free draining.

4. Figure 96 shows a proposed embankment in sandy clay for which the angle of shearing resistance is 10°, the cohesion is 11·3 kN/m^2 and the bulk density is 1840 kg/m^3. Find the factor of safety against a toe failure along a circular slip surface tangential to bedrock and with its centre on AB, assuming a pore-pressure coefficient of 0·5 and that no tension cracks will occur.

5. (a) The arrangement of a timber berth on a canal is shown in Fig. 97. If tension cracks have not yet formed, determine the factor of safety against a slip along the circle shown. The loading from the berth inclusive of its dead weight may be taken as 50 kN/m^2 of platform area. The soil is saturated throughout, its saturated density being 1980 kg/m^3 and its shear strength 50 kN/m^2 ($\varphi = 0°$). G.W.L. is the same as the water level in the canal.

(b) What would be the new factor of safety if the canal bed were dredged out a further 1·5 m as shown?

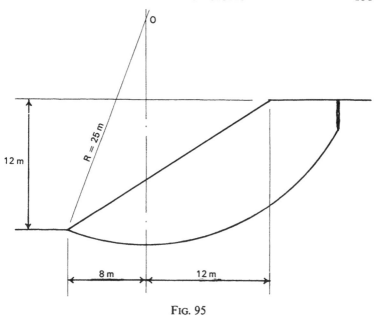

FIG. 95

6. An embankment is to be cut as shown in Fig. 98. Calculate the factor of safety of the embankment associated with a possible circular plane of failure, which has its centre located at O. The density of silt is to be taken as 1920 kg/m^3 and clay 1840 kg/m^3. Neglect the effect of tension cracks.

7. A temporary excavation is to be made in a clay soil 10 metres deep with a side slope of 1:1 as shown in Fig. 99. Investigate the factor of safety against failure if the critical slip circle is assumed tangential to the sandstone stratum shown and passes through the toe of the slope. The centre of the circle may be assumed to lie on a vertical line through the top of the slope. The formation of tension cracks may be ignored and reasonable estimates of areas and centroids made.

A series of undrained tri-axial tests gave parameters c_u and φ_u as 30 kN/m^2 and 0° respectively, the bulk density of the soil is 2000 kg/m^3.

Comment on the factor of safety calculated and suggest any modifications or specifications which would improve the stability stating the reason.

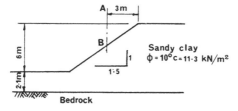

FIG. 96

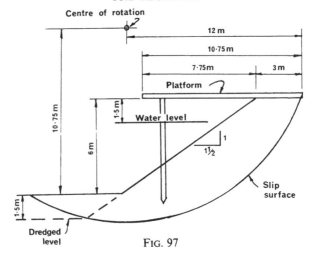

FIG. 97

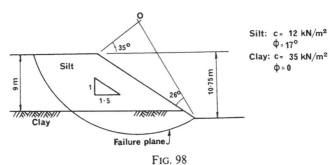

FIG. 98

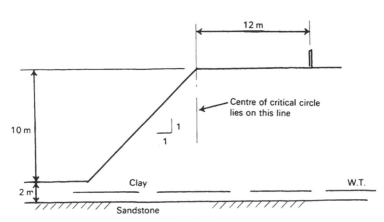

FIG. 99

8. Figure 100 shows a possible slip circle, centre O, in a temporary cutting. Materials are to be stored at the toe of the slope giving a uniformly distributed load of $100 \, kN/m^2$. Calculate the factor of safety for this slip circle. To obtain a clear working space at formation level, the contractor moved his materials to the top of the slope into the position shown by a broken line in the figure. How does this affect the stability of the slope? Assume tension cracks do not form.

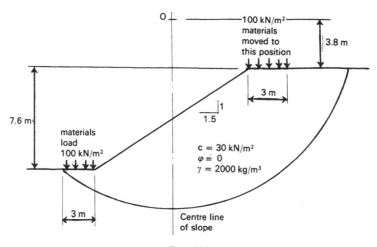

FIG. 100

CHAPTER 10

FOUNDATIONS

The most common cause of foundations failure is excessive or differential settlement. The estimation of settlement is dealt with in
Chapter 5.

The ultimate bearing capacity of soils, however, is based on the
shear strength of the soil, but it should be remembered this is rarely
the criterion in final design calculations.

SHALLOW FOUNDATIONS

A number of analyses have been carried out to find the ultimate
bearing capacity of the soil, q_u, when the footing is at the surface as
shown in Fig. 101

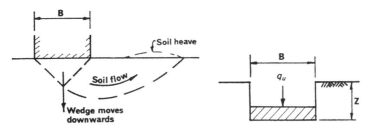

FIG. 101. Rupture zone theory   FIG. 102. Footing below ground level

Terzaghi carried out an investigation in which he considered the
footing below the surface and allowed for friction and cohesion
between the footing and the foundation. The following equation was
derived (*see* Fig. 102):

For strip footings:

$$q_u = cN_c + \gamma z(N_q - 1) + 0.5\gamma BN_y$$

N_c, N_q and N_y are the bearing capacity coefficients which may be
obtained from Fig. 103 and depend on φ.

The first term cN_c, deals with the cohesion. If z is taken as 0 (i.e.
footing at the surface) and φ is 0, then $N_y = 0$, $(N_q - 1) = 0$ and
$N_c = 5.7$; i.e. $q_u = 5.7c$.

The third term, $0.5\ \gamma BN_y$, applies only to frictional soils ($N_y = 0$

when $\varphi = 0$) and is the term which takes into account the breadth of footing.

The second term, $\gamma z(N_q - 1)$, takes into account the load due to overburden. If $(N_q - 1)$ is used the term q_u may be increased by the weight of soil removed, γz. Sometimes this term is taken as $\gamma z N_q$, in which case any factor of safety applied would also be applied to the overburden.

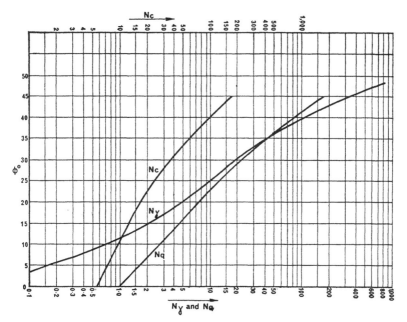

FIG. 103. Bearing capacity coefficients

If a rectangular footing width B, length L is being used then friction at the ends of the footing must also be considered. The only variation will be to the bearing capacity coefficients N_c and N_y which are multiplied by the following factors:

N_c (for rectangular footing)

$$= N_c \text{ (for strip footing)} \times \left(1 + \frac{0.2B}{L}\right)$$

N_y (for rectangular footing)

$$= N_y \text{ (for strip footing)} \times \left(1 - \frac{0.2B}{L}\right)$$

Coefficient N_q will be unchanged.

With these corrections to bearing capacity coefficients the equation for a strip footing may then be used.

EXAMPLE 40

Terzaghi's formula for the net ultimate bearing capacity q_u (total pressure less overburden pressure) for a strip footing is:

$$q_u = cN_c + \gamma z(N_q - 1) + \tfrac{1}{2}\gamma BN_\gamma$$

For a certain soil the cohesion c is 48 kN/m^2, the density γ is 1930 kg/m^2 and the coefficients are $N_c = 8$, $N_q = 3$ and $N_\gamma = 2$. Calculate the net ultimate bearing capacity for a strip footing of width $B = 2$ m at a depth $z = 1$ m.

Considering shear failure only, calculate the safe total load on a footing 6 m long by 2 m wide, using a load factor of 2·5.

What other soil properties should be taken into account in determining the safe load on this foundation?

SOLUTION

Net $q_u = cN_c + \gamma z(N_q - 1) + \tfrac{1}{2}\gamma BN_\gamma$
$ = 48 \times 8 + 1·93 \times 9·8 \times 1 \times 2 + \tfrac{1}{2} \times 1·93 \times 9·8 \times 2 \times 2$
$ = \underline{\underline{497 \text{ kN/m}^2}}$

$$Q = \left(\frac{497}{2·5} + 1 \times 1·93 \times 9·8\right)6 \times 2$$
$$= \underline{\underline{2612 \text{ kN}}}$$

(*Note:* The load factor is not applied to $1 \times 1·93 \times 9·8$.)

Consolidation and settlement should also be considered, which would include compressibility and permeability of the soil.

Another method of studying the stability of foundations is by assuming a slip circle rotating about one edge of the footing.

EXAMPLE 41

(a) Derive an expression for the ultimate bearing capacity of a cohesive soil under a long strip footing, of width B and depth z, on the assumption that failure will take place by rotation about one edge of the footing.

(b) Compare the formula obtained with Terzaghi's formula:

$$q_u = cN_c + \gamma z N_q + \tfrac{1}{2}\gamma BN_\gamma$$

(c) Describe and compare the ways in which the *safe* bearing capacity may be found from the Terzaghi formula.

SOLUTION

(a) Consider a slip circle under the footing, as shown in Fig. 104.

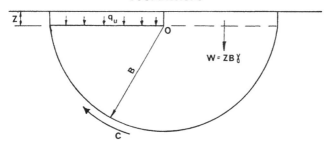

FIG. 104. Bearing capacity: slip circle

For unit length of long strip footing:
Disturbing moment about O:

$$Bq_u \times \frac{B}{2} = \frac{q_u B^2}{2}$$

Resisting moments about O:

(i) *Cohesion:* $\pi Bc \times B + zc \times B = \pi B^2 c + zcB$

(ii) *Gravity:* $zB\gamma \times \dfrac{B}{2} = \dfrac{z\gamma B^2}{2}$

For stability the disturbing moment and resisting moment must be equal

$$\frac{q_u B^2}{2} = \pi B^2 c + zcB + \frac{z\gamma B^2}{2}$$

$$q_u = 2\pi c + \frac{2zc}{B} + z\gamma$$

or $$q_u = c(2\pi) + \gamma z + \frac{2zc}{B}$$

(b) The above proof is suitable for cohesive soils only where $\varphi = 0$. When $\varphi = 0$ Terzaghi's formula gives $q_u = 5 \cdot 7c + \gamma z$, which is comparable.

(c) Referring to Terzaghi's formula, the safety factor F_s may be applied two ways, thus:

$$q_{safe} = \frac{5 \cdot 7c}{F_s} + \gamma z \quad \text{or} \quad q_{safe} = \frac{5 \cdot 7c + \gamma z}{F_s}$$

Which of these two methods is used will depend upon the second term of the equation:

$$\gamma z(N_q - 1) \quad \text{or} \quad \gamma z N_q$$

DEEP FOUNDATIONS

Foundations are considered deep where the structure has a deep basement or where it is supported on piles.

Deep basements are designed in a similar way to raft foundations, but different values of the bearing capacity coefficients are used. Care must also be taken to ensure the load imposed by the building is sufficient to prevent uplift.

A piled foundation transfers the load to more resistant strata at depth. Piles may be pre-formed and driven into the ground or a hole bored and filled with concrete. Driven piles cause displacement of the surrounding soil and so are referred to as large displacement piles. Bored piles are referred to as non-displacement piles. There are many piling techniques available, but the basic design procedure is common to most methods.

Piles in clay

The load-carrying capacity of a pile depends on two factors, the bearing pressure at the bottom of the pile and the adhesion or friction between the surface of a pile along its length and the surrounding soil. Generally one or other of these two factors will be the criterion of pile bearing capacity but both factors may be considered.

End bearing

The bottom of the pile will bear on a stratum with a shear strength c_{ub}. For deep foundations the bearing capacity coefficient N_c is greater than for shallow foundations and is normally taken as 9. For a pile with a base cross-sectional area A_b therefore, the end bearing capacity will be $9c_{ub}A_b$.

Bored piles may have the base reamed out to a larger diameter than the shaft in order to give a higher bearing capacity.

Skin adhesion

Skin adhesion or, as it is more commonly called skin friction, is the load transmitted to the soil surrounding the shaft of the pile. If the average value of the shear strength of the soil along the length of the pile is c_{ua} and the surface area of the shaft A_s, then the maximum possible value would be $c_{ua}A_s$. However, it is not possible for the adhesion between the pile and surrounding soil to reach this value, and an adhesion factor α must be applied. The value of the adhesion factor α is a subject of considerable research, but the most commonly accepted value is $\alpha = 0.45$.

A possible problem with skin friction is that the surrounding soil may consolidate after the pile is in place, dragging the pile down with it. If this occurs the skin friction will become negative and must be *deducted* from the bearing capacity of the pile.

When piles are driven in clay the pore water at the pile–soil interface is disturbed, and it may be several weeks before equilibrium is restored. For this reason piles driven in clay should not be subjected to load immediately after driving.

EXAMPLE 42
Figure 105 shows details of a bore-hole log on a site. 500 mm diameter bored piles 15 m long are specified. Estimate the safe bearing capacity of one pile assuming the loose fill will settle after the piles are placed. Take a factor of safety of 2·5

Compare the bearing capacity of a 500-mm-diameter pile 12 m long and of uniform cross section, with a similar pile which has had the bottom 2 m reamed out to 900 mm diameter.

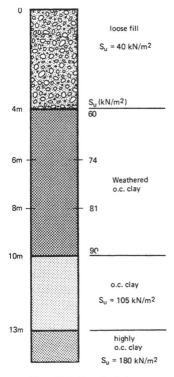

FIG. 105. Borehole log

SOLUTION
For 500 mm diameter pile 15 m long:

$$\text{End bearing} = 9 \times 180 \times \frac{\pi \times 0.5^2}{4} = 318 \text{ kN}$$

Skin adhesion:

Weathered over-consolidated clay
$$= 6 \times \pi \times 0{\cdot}5 \left(\frac{60 + 74 + 81 + 90}{4} \right) \times 0{\cdot}45$$

$$= 323 \text{ kN}$$
$$\text{o.c. clay} = 3 \times \pi \times 0{\cdot}5 \times 105 \times 0{\cdot}45 = 223 \text{ kN}$$
$$\text{Highly o.c. clay} = 2 \times \pi \times 0{\cdot}5 \times 180 \times 0{\cdot}45 = 254 \text{ kN}$$
$$\text{Load carrying capacity} = \frac{318 + 323 + 223 + 254}{2{\cdot}5} = 447 \text{ kN}$$

$$\text{Negative skin friction} = 4 \times \pi \times 0{\cdot}5 \times 40 \times 0{\cdot}45 = 113 \text{ kN}$$
$$\text{Safe bearing capacity} = 447 - 113 = \underline{\underline{334 \text{ kN}}}$$

For 500 mm diameter pile 10 m long with belled out base:

$$\text{End bearing} = 9 \times 105 \times \frac{\pi \times 0{\cdot}9^2}{4} = 601 \text{ kN}$$

Skin adhesion:
$$\text{Weathered o.c. clay} = 323 \text{ kN}$$
(skin adhesion cannot be taken on the under-reamed section of the pile)

$$\text{Load carrying capacity} = \frac{601 + 323}{2{\cdot}5} = 370 \text{ kN}$$

$$\text{Negative skin friction} = 113 \text{ kN}$$
$$\text{Safe bearing capacity} = \underline{\underline{257 \text{ kN}}}$$

Piles in sand
A similar exercise may be carried out for piles in sand, i.e. calculate end bearing and skin friction in order to estimate pile-bearing capacity. In sand, however, end bearing will generally be much greater, since the relevant term in the bearing capacity equation is $\gamma z (N_q - 1)$ and for sand with values of φ above 30°, N_γ for deep foundations will have a value greater than 60. Skin friction is a more applicable term in sands, and is based on the lateral pressure multiplied by a coefficient of friction.

Pile driving formulae
An alternative method of design for driven piles is to use a pile driving formula. The basic assumption is that the energy transmitted to the pile by the hammer is equal to the energy absorbed by the pile in penetrating the soil.

i.e. $W \times h = s \times Q$

where: W = weight of pile driving hammer;
       h = height through which hammer falls;
       s = penetration of pile per blow (set);
       Q = soil resistance overcome.

The basic assumption that no energy loss occurs is, of course, inaccurate, and most pile driving formulae apply some factor to allow for energy lost due to the inefficiency of the hammer, compression of the pile, etc.

i.e. $$W \times h \times \eta = s \times Q$$

where: η = efficiency coefficient.

EXAMPLE 43

A pile is to be driven with a hammer weighing 200 kg with a drop of 2 m. If the pile is required to carry a load of 1500 kN, what set would be specified for 10 blows of the hammer? Assume that the efficiency of energy transmission is 60 per cent.

SOLUTION

$$\frac{200 \times 9.8}{1000} \times 2 \times 0.6 = s \times 1500$$

$$s = \frac{200 \times 9.8 \times 2 \times 0.6}{1000 \times 1500} \times 1000 \text{ mm per blow}$$

$$= 1.6 \text{ mm per blow}$$

For ten blows a maximum set of 16 mm could be specified.

Alternative methods of estimating the carrying capacity of piles in sand are based on cone or standard penetration tests. It should be remembered, however, that all pile bearing capacity formulae are open to error, and load tests should always be carried out on a selection of piles to check their authenticity.

Groups of piles

When piles are driven close together, the bearing capacity of the group of piles may not be the sum of the bearing capacities of the individual piles.

In sands, if the piles are closer than six times the diameter (or breadth) of the pile, the sand tends to compact and the bearing capacity of the group will be greater than the sum of the bearing capacities of the individual piles.

In clays, closely driven piles will tend to reduce the shear strength of the clay, and hence the bearing capacity of the group may be less than the sum of the bearing capacities of individual piles.

QUESTIONS

1. A structure is to be erected on a flat site, over most of which the soil is a partially saturated silty clay with $\varphi_u = 20°$, $c_u = 5 \text{ kN/m}^2$ and bulk density $\gamma = 1920 \text{ kg/m}^3$. Areas occur in which the material is predominantly clay, having $\varphi_u = 0°$, $c_u = 50 \text{ kN/m}^2$. The water table is some distance

below ground level. It is proposed to install strip footings at a depth of 1 m below ground level.

Determine the width of footing required on each type of soil if the loading intensity (including an estimate of the weight of the footings) is 120 kN/m. A load factor of 3 against shear failure is required on both soils.

2. A square footing is required to carry a total load of 3 MN at a depth $z = 3$ m on soil of density $\gamma = 1920$ kg/m³. The properties of the soil as found from undrained shear tests are: cohesion $c = 100$ kN/m² and angle of shearing resistance $\varphi = 0°$. Find the breadth B of the footing, using the undrained soil properties with a factor of safety of 3 against shear failure.

3. A clay stratum extending to considerable depth has the following properties: $\varphi_u = 0$, c_u – varies linearly from 100 kN/m² at the surface to 250 kN/m² at 20 m depth. Bulk density of clay = 2000 kg/m³.

(a) Estimate the allowable bearing capacity of a bored pile 800 mm diameter and 16 m long, with a factor of 2·5.

(b) Estimate the allowable bearing capacity for a strip footing 1 metre deep in this clay. Use a factor of safety of 3.

(c) Assuming the bearing capacity determined in (a) is specified but the

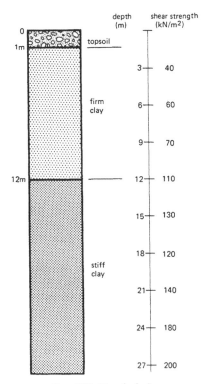

FIG. 106. Borehole log

piles are to be driven with a 2 kN hammer falling 3 m, what set would you specify to the contractor. Assume 75 per cent efficiency for pile driving equipment.

4. Figure 106 shows the borehole log for a site in which bored piles 1000 mm diameter are to be placed to a depth of 24 m. Estimate the load-carrying capacity of the piles, taking a factor of safety of 2·5.

5. A foundation consists of 70 piles each of length 20 m and diameter 750 mm. The piles form a group which supports a total load of 45×10^3 kN. The tops of the piles are located at a depth of 1·6 m in a clay stratum in which $\varphi = 0$ and c varies uniformly from 54 kN/m² at the ground surface to 220 kN/m² at a depth of 30 m.

Given that the adhesion factor α is 0·45, the bearing capacity factor N_c is 9 and the efficiency of the pile group is 70 per cent calculate the factor of safety for the group.

APPENDIX

ANSWERS TO QUESTIONS

(*Note:* Where questions require graphs, flow nets or scale diagrams, answers can only be approximate.)

Chapter 1
1. $\gamma_d = 1496$ kg/m^3, $e = 0.767$, $G_s = 2.64$; $\gamma = 1887$ kg/m^3
2. (a) $\gamma_d = 1558$ kg/m^3; (b) $e = 0.745$; (c) $V_a = 10.7\%$. At saturation γ_{sat} = 1985 kg/m^3, $m = 27.4\%$
3. $\gamma = 1954$ kg/m^3, $\gamma_d = 1566$ kg/m^3, $m = 24.7\%$, at saturation $m = 26.9\%$
4. (a) $G_s = 2.72$; (b) $m = 20.1\%$; (c) $e = 0.887$; (d) $\gamma = 1730$ kg/m^3, $\gamma_d = 1440$ kg/m^3, $\gamma_{sat} = 1910$ kg/m^3, $\gamma' = 910$ kg/m^3; (e) $i_c = 0.91$
5. 5. (i) 104·4 kN/m^2; (ii) 124 kN/m^2; (iii) 84·8 kN/m^2; (iv) 65·2 kN/m^2; (v) 65·2 kN/m^2

Chapter 2
1. o.m.c. = 10%, max. comp = 2002 kg/m^3 $e = 0.35$, $n = 0.26$, $S_r = 77\%$
2. (c) o.m.c. = 11%, max. comp. = 2012 kg/m^3; (e) $V_a = 2\%$; (g) $m = 11.9\%$
3. $\sigma = 352$ kN/m^2, $e = 0.66$, $S_r = 48\%$

Chapter 3
1. Poorly graded sandy gravel, $D_{10} = 0.48$, $U = 9.3$
2. 4·3 hours
3. Clay of high plasticity (CH)

Chapter 4
1. 11.8 ml
2. $k_H = 0.72$ mm/s, $k_V = 0.3$ mm/s
3. 0·15 mm/s
4. $k = 4.4 \times 10^{-3}$ mm/s, $Q = 42\,000$ l/day/10 m run
5. 740 l/day/m run
6. 5400 l/day/m run
8. 140 mm/hour, F. of S. = 2·4
9. Reduction in seepage 37 000 l/day, change in F. of S. = 67% increase

Chapter 5
1. $c_v = 0.004$ m^2/day
2. $\rho = 195$ mm $t_{90} = 3.6$ years
3. $p_0' = 59.8$ kN/m^2, $p_1' = 132.7$ kN/m^2, $\rho = 48$ mm
4. $m_v = 0.28 \times 10^{-3}$ m^2/kN, $k = 30 \times 10^{-6}$ mm/s
5. 9 months, $k = 2.2 \times 10^{-7}$ mm/s

Chapter 6
 1. $c = 47 \text{ kN/m}^2$, $\varphi = 16°$, $\sigma_1 = 300 \text{ kN/m}^2$
 2. $c = 69 \text{ kN/m}^2$
 3. $\tau = 60·6 \text{ kN/m}^2$
 4. (a) $c = 30 \text{ kN/m}^2$, $\varphi = 15°$; (c) $\tau = 62 \text{ kN/m}^2$
 5. $c' = 10$, $\varphi' = 22°$, $u = 155 \text{ kN/m}^2$
 6. $c = 116 \text{ kN/m}^2$, $c_v/c^\theta = 1·3$
 7. $R = 0·5$
 8. $\Delta u = 37 \text{ kN/m}^2$

Chapter 7
 3. 20 kN/m^2
 5. Point 1: 86 kN/m², Point 2: 93 kN/m
 6. (a) $\sigma_{VA} = 47 \text{ kN/m}^2$, $\sigma_{VB} = 16 \text{ kN/m}^2$ (b) $\sigma_{VC} = 32 \text{ kN/m}^2$, $\Delta\rho = 7$ mm

Chapter 8
 1. (a) $P_a = 692 \text{ kN}$; (b) $z_0 = 2·7 \text{ m}$; (c) $h = 3 \text{ m}$
 2. (a) $P = 86 \text{ kN}$ acting 1·4 m above base; (b) 52 kN
 3. Approx. 500 kN/m
 4. Approx. 260 kN/m
 5. Min $P_p = 205 \text{ kN}$, F. of S. = 1·48

Chapter 9
 1. F. of S. = 1·9
 2. F. of S. = 2
 3. F. of S. = 1·9
 4. F. of S. = 2·2
 5. (a) F. of S. = 2·2; (b) F. of S. = 1·8
 6. F. of S. = 1·55
 7. F. of S. = 1·1
 8. F. of S. = 1·5. Load moved F. of S. = 0·98

Chapter 10
 1. Silty clay $B = 1·15 \text{ m}$, clay $B = 1·18 \text{ m}$
 2. $B = 3·25 \text{ m}$
 3. $Q_a =$ (i) 1556 kN, (ii) 224 kN/m² (iii) 50 mm/10 blows
 4. $Q_a = 1776 \text{ kN}$
 5. F. of S. = 3·48

INDEX

Printed and bound by CPI Group (UK) Ltd, Croydon, CR0 4YY

01/11/2024

01782614-0002